Selected Titles in This Series

D1805564

735 **Martina Brück, Xi Du, Joonsang Park, and Chuu-Lian Terng,** The submanifold geometries associated to Grassmannian systems, 2002

734 **Michel Van den Bergh,** Blowing up of non-commutative smooth surfaces, 2001

733 **Milé Krajčevski,** Tilings of the plane, hyperbolic groups and small cancellation conditions, 2001

732 **Jan O. Kleppe, Juan C. Migliore, Rosa Miró-Roig, Uwe Nagel, and Chris Peterson,** Gorenstein liaison, complete intersection liaison invariants and unobstructedness, 2001

731 **Jesús Bastero, Mario Milman, and Francisco J. Ruiz,** On the connection between weighted norm inequalities, commutators and real interpolation, 2001

730 **Suhyoung Choi,** The decomposition and classification of radiant affine 3-manifolds, 2001

729 **Michael Grosser, Eva Farkas, Michael Kunzinger, and Roland Steinbauer,** On the foundations of nonlinear generalized functions I and II, 2001

728 **Laura Smithies,** Equivariant analytic localization of group representations, 2001

727 **Anthony D. Blaom,** A geometric setting for Hamiltonian perturbation theory, 2001

726 **Victor L. Shapiro,** Singular quasilinearity and higher eigenvalues, 2001

725 **Jean-Pierre Rosay and Edgar Lee Stout,** Strong boundary values, analytic functionals, and nonlinear Paley-Wiener theory, 2001

724 **Lisa Carbone,** Non-uniform lattices on uniform trees, 2001

723 **Deborah M. King and John B. Strantzen,** Maximum entropy of cycles of even period, 2001

722 **Hernán Cendra, Jerrold E. Marsden, and Tudor S. Ratiu,** Lagrangian reduction by stages, 2001

721 **Ingrid C. Bauer,** Surfaces with $K^2 = 7$ and $p_g = 4$, 2001

720 **Palle E. T. Jorgensen,** Ruelle operators: Functions which are harmonic with respect to a transfer operator, 2001

719 **Steve Hofmann and John L. Lewis,** The Dirichlet problem for parabolic operators with singular drift terms, 2001

718 **Bernhard Lani-Wayda,** Wandering solutions of delay equations with sine-like feedback, 2001

717 **Ron Brown,** Frobenius groups and classical maximal orders, 2001

716 **John H. Palmieri,** Stable homotopy over the Steenrod algebra, 2001

715 **W. N. Everitt and L. Markus,** Multi-interval linear ordinary boundary value problems and complex symplectic algebra, 2001

714 **Earl Berkson, Jean Bourgain, and Aleksander Pełczynski,** Canonical Sobolev projections of weak type $(1, 1)$, 2001

713 **Dorina Mitrea, Marius Mitrea, and Michael Taylor,** Layer potentials, the Hodge Laplacian, and global boundary problems in nonsmooth Riemannian manifolds, 2001

712 **Raúl E. Curto and Woo Young Lee,** Joint hyponormality of Toeplitz pairs, 2001

711 **V. G. Kac, C. Martinez, and E. Zelmanov,** Graded simple Jordan superalgebras of growth one, 2001

710 **Brian Marcus and Selim Tuncel,** Resolving Markov chains onto Bernoulli shifts via positive polynomials, 2001

709 **B. V. Rajarama Bhat,** Cocylces of CCR flows, 2001

708 **William M. Kantor and Ákos Seress,** Black box classical groups, 2001

707 **Henning Krause,** The spectrum of a module category, 2001

706 **Jonathan Brundan, Richard Dipper, and Alexander Kleshchev,** Quantum Linear groups and representations of $GL_n(\mathbb{F}_q)$, 2001

705 **I. Moerdijk and J. J. C. Vermeulen,** Proper maps of toposes, 2000

(Continued in the back of this publication)

The Submanifold Geometries Associated to Grassmannian Systems

MEMOIRS
of the
American Mathematical Society

Number 735

The Submanifold Geometries Associated to Grassmannian Systems

Martina Brück
Xi Du
Joonsang Park
Chuu-Lian Terng

January 2002 • Volume 155 • Number 735 (first of 5 numbers) • ISSN 0065-9266

American Mathematical Society
Providence, Rhode Island

2000 *Mathematics Subject Classification.*
Primary 53–XX, 35–XX.

Library of Congress Cataloging-in-Publication Data

The submanifold geometries associated to Grassmannian systems / Martina Brück . . . [et al.].
p. cm. — (Memoirs of the American Mathematical Society, ISSN 0065-9266 ; no. 735)
"January 2002."
"Volume 155, number 735 (first of 5 numbers)."
Includes bibliographical references.
ISBN 0-8218-2753-7 (alk. paper)
1. Grassmann manifolds. 2. Submanifolds. I. Brück, Martina, 1970– II. Series.

QA3 .A57 no. 735
[QA613.6]
510 s—dc21
[516.3′6] 2001045782

Memoirs of the American Mathematical Society

This journal is devoted entirely to research in pure and applied mathematics.

Subscription information. The 2002 subscription begins with volume 155 and consists of six mailings, each containing one or more numbers. Subscription prices for 2002 are $524 list, $419 institutional member. A late charge of 10% of the subscription price will be imposed on orders received from nonmembers after January 1 of the subscription year. Subscribers outside the United States and India must pay a postage surcharge of $31; subscribers in India must pay a postage surcharge of $43. Expedited delivery to destinations in North America $35; elsewhere $130. Each number may be ordered separately; *please specify number* when ordering an individual number. For prices and titles of recently released numbers, see the New Publications sections of the *Notices of the American Mathematical Society.*

Back number information. For back issues see the *AMS Catalog of Publications.*

Subscriptions and orders should be addressed to the American Mathematical Society, P. O. Box 845904, Boston, MA 02284-5904. *All orders must be accompanied by payment.* Other correspondence should be addressed to Box 6248, Providence, RI 02940-6248.

Memoirs of the American Mathematical Society is published bimonthly (each volume consisting usually of more than one number) by the American Mathematical Society at 201 Charles Street, Providence, RI 02904-2294. Periodicals postage paid at Providence, RI. Postmaster: Send address changes to Memoirs, American Mathematical Society, P. O. Box 6248, Providence, RI 02940-6248.

This publication is indexed in *Science Citation Index®*, *SciSearch®*, *Research Alert®*, *CompuMath Citation Index®*, *Current Contents®/Physical, Chemical & Earth Sciences.*
Printed in the United States of America.

∞ The paper used in this book is acid-free and falls within the guidelines established to ensure permanence and durability.
Visit the AMS home page at URL: `http://www.ams.org/`

10 9 8 7 6 5 4 3 2 1 07 06 05 04 03 02

Contents

1. Introduction 1

2. The U/K-system 7

3. $G_{m,n}$-systems 12

4. $G^1_{m,n}$-systems 18

5. Moving frame method for submanifolds 21

6. Submanifolds associated to $G_{m,n}$-systems 25

7. Submanifolds associated to $G^1_{m,n}$-systems 34

8. $G^1_{m,1}$-systems and isothermic surfaces 39

9. Loop group action for $G_{m,n}$-systems 45

10. Ribaucour transformations for $G_{m,n}$-systems 54

11. Loop group actions for $G^1_{m,n}$-systems 63

12. Ribaucour transformations for $G^1_{m,n}$-systems 67

13. Darboux transformations for $G^1_{m,1}$-systems 68

14. Bäcklund transformations and loop group factorizations 73

15. Permutability formula for Ribaucour transformations 80

16. The U/K-hierarchy and finite type solutions 83

Pictures 87

Bibliography 94

Abstract

Some high points in classical differential geometry are: 1) to study surfaces in R^3 with special geometric properties, 2) to find good coordinates so that the corresponding Gauss-Codazzi equations have specially nice forms, 3) to construct explicit examples and deformations of such surfaces. Surfaces with negative constant Gaussian curvature, surfaces with constant mean curvature, and isothermic surfaces are some of the well-known examples. The Gauss and Codazzi equations of these surfaces are now known to be "soliton" equations. In recent years, modern geometers have found that these equations admit "Lax pairs", i.e., they can be written as the condition for a family of connections to be flat. The existence of a Lax pair is one of the characteristic properties of soliton equations, and it often gives rise to an action of an infinite dimensional loop group on the space of solutions (the dressing action). The geometric transformations found for these surfaces by classical geometers, such as Bäcklund, Darboux, and Ribaucour, often arise as the dressing action of some simple rational elements in the loop group. In this approach, one often starts with a class of surfaces in R^3, and if there are methods to construct an infinite parameter family of solutions from a given one, then it hints that we may be able to find a good coordinate system and a Lax pair. In this monograph, we use a different approach. We consider well-known soliton equations and study the geometries associated to them. It is known that we can associate to each symmetric space U/K a hierarchy of soliton equations. For example, the $SU(2)$-hierarchy is the one for the non-linear Schrödinger equation and the $SU(2)/SO(2)$-hierarchy is the one associated to the modified KdV equation. If the rank of the symmetric space U/K is n, then the n first flows give rise to a natural non-linear first order system, the U/K-system. The main goal of this monograph is to study submanifold geometries associated to $O(m+n)/O(m) \times O(n)$ and $O(m+n,1)/O(m) \times O(n,1)$-systems. These include submanifolds with constant sectional curvatures in space forms, isothermic surfaces, and submanifolds admitting special principal curvature coordinates. We also develop a systematic approach for associating submanifold geometries to given soliton equations.

2000 *Mathematics Subject Classification.* 53, 35

Keywords and Phases. Soliton equations, submanifolds, geometric transformations

1. Introduction

Some high points in classical differential geometry are: to study surfaces in R^3 with special geometric properties, to find good coordinates so that the corresponding Gauss-Codazzi equations have specially nice forms, and to construct explicit examples and deformations of these surfaces. Surfaces with negative constant Gaussian curvature, surfaces with constant mean curvature, and isothermic surfaces are some of the well-known examples. The Gauss and Codazzi equations of these surfaces are now known to be "soliton" equations. In recent years, modern geometers have found that these equations admit "Lax pairs", i.e., they can be written as the condition for a family of connections to be flat. The existence of a Lax pair is one of the characteristic properties of soliton equations, and it often gives rise to an action of an infinite dimensional group on the space of solutions (the dressing action). The geometric transformations found for these surfaces by classical geometers such as Bäcklund, Darboux, and Ribaucour, often arise as the dressing action of some simple rational elements. For more detail see [Bo2], [Bo3], [TU3]. In this approach, one starts with a class of surfaces in R^3, and if there are methods to construct an infinite parameter family of solutions from a given one, then it hints that we may be able to find a good coordinate system and a Lax pair. Geometers have used this method to construct soliton equations involving n variables (cf. [TT], [Te1], [FP2]). But there is no uniform algorithm to achieve this or determine whether a geometric equation for a certain class of submanifolds is a soliton equation.

It is also known that we can associate to each symmetric space U/K a hierarchy of soliton equations (cf. [TU1]). For example, the $SU(2)$-hierarchy is the hierarchy for the non-linear Schrödinger equation and the $SU(2)/SO(2)$-hierarchy is the hierarchy for the modified KdV equation. If the rank of the symmetric space U/K is n, Terng [Te2] put the n first flows together to construct a natural non-linear first order system, the U/K-system, and initiated the project of identifying the submanifold geometry associated to these systems. This means to find submanifolds in certain symmetric space M whose Gauss-Codazzi equation is given by the U/K-system and to find the geometric transformations corresponding to the dressing actions of certain simple elements. This direct approach may provide ways to find Lax pairs for some known class of submanifolds, and also may give new interesting class of submanifolds. The main goal of this paper is to carry out this project for the real Grassmannian manifolds of space-like m-dimensional linear subspaces in R^{m+n} and in $R^{m+n,1}$.

Below we give a short review of some known facts and outline our results.

- **The U/K-system**

Let U be a semi-simple Lie group, σ an involution on U, and K the fixed point set of σ. Then U/K is a symmetric space. The Lie algebra $\mathcal{K}$ is the $+1$ eigenspace

---

Received by the editor June 15, 2000.

Xi Du, Martina Brück and Chuu-Lian Terng's research were supported in part by NSF grant DMS 9626130.

Joonsang Park's research was supported by grant No. 2000-2-10100-002-3 from the Basic Research Program of the Korea Science & Engineering Foundation.

of the differential σ_* of σ at the identity. Let $\mathcal{P}$ denote the -1 eigenspace of σ_*. Then $\mathcal{U} = \mathcal{K} \oplus \mathcal{P}$ and

$$[\mathcal{K}, \mathcal{K}] \subset \mathcal{K}, \quad [\mathcal{K}, \mathcal{P}] \subset \mathcal{P}, \quad [\mathcal{P}, \mathcal{P}] \subset \mathcal{P}.$$

Let $\mathcal{A}$ be a maximal abelian subalgebra in $\mathcal{P}$, $a_1, \ldots, a_n$ a basis for $\mathcal{A}$, and $\mathcal{A}^\perp$ the orthogonal complement of $\mathcal{A}$ in $\mathcal{U}$ with respect to the Killing form $<,>$. We recall that $n = \dim(\mathcal{A})$ is called the *rank* of the symmetric space. *The n-dimensional system associated to U/K*, defined by Terng in [Te2], is the following first order non-linear partial differential equation for $v : R^n \to \mathcal{P} \cap \mathcal{A}^\perp$:

$$[a_i, v_{x_j}] - [a_j, v_{x_i}] = \big[[a_i, v], [a_j, v] \big], \qquad 1 \leq i \neq j \leq n, \tag{1.1}$$

where $v_{x_i} = \frac{\partial v}{\partial x_i}$. We will call such system the *U/K-system*.

- **The Lax connection**

Recall that a $\mathcal{G}$-valued connection $\frac{\partial}{\partial x_i} + A_i$ is flat if its curvature is zero, i.e.,

$$\left[\frac{\partial}{\partial x_i} + A_i, \ \frac{\partial}{\partial x_j} + A_j \right] = 0$$

for all i, j. Let V be a linear subspace of a Lie algebra $\mathcal{G}$. A partial differential equation (PDE) for $v : R^n \to V$ admits a *Lax connection* if there exists a family of $\mathcal{G}$-valued connections

$$\frac{\partial}{\partial x_i} + A_i(v, dv, d^2 v, \cdots, d^k v, \lambda)$$

such that the PDE for v is given by the flatness of these connections for all λ in some open domain in C. Equation (1.1) admits a Lax connection, $\frac{\partial}{\partial x_i} + a_i \lambda + [a_i, v]$. In other words, v is a solution of (1.1) if and only if

$$\left[\frac{\partial}{\partial x_i} + a_i \lambda + [a_i, v], \ \frac{\partial}{\partial x_j} + a_j \lambda + [a_j, v] \right] = 0 \tag{1.2}$$

for all i, j and $\lambda \in C$. When $n = 2$, a Lax connection gives rise to a pair of commuting operators. This was first observed by Lax for the KdV equation and is called a *Lax pair* in the soliton literature.

Note that the connection $\frac{\partial}{\partial x_i} + A_i$ is flat if and only if the connection 1-form $\omega = \sum_i A_i dx_i$ is flat, i.e., $d\omega = -\omega \wedge \omega$. In particular, v is a solution of the U/K-system (1.1) if and only if

$$\theta_\lambda = \sum_{i=1}^{n} (a_i \lambda + [a_i, v]) dx_i \tag{1.3}$$

is flat for all $\lambda \in C$. We will also call θ_λ the Lax connection of the U/K-system.

• **Dressing action**

The existence of a Lax connection for an equation often gives rise to an action of certain subgroup of germs of holomorphic maps at a suitable point on the space of local solutions of the equation. This is called "dressing action" in the soliton literature. Below we give a rough sketch of the construction of the dressing action (cf. [ZS], [Ch], [TU2]).

If v is a solution of (1.1), then its Lax connection θ_λ is flat for all $\lambda \in C$, so there exists a U_C-valued map $E(x,\lambda)$ such that

$$E^{-1}dE = \theta_\lambda, \quad E(0,\lambda) = I. \tag{1.4}$$

Since θ_λ is holomorphic for $\lambda \in C$, so is $E(x,\lambda)$. Now let $g(\lambda)$ be a holomorphic map defined from a neighborhood of $\lambda = \infty$ in $S^2 = C \cup \{\infty\}$ to U_C that satisfies certain U/K-reality condition (defined later) and $g(\infty) = I$. It follows from the classical Birkhoff factorization theorem (cf. [PS]) that there exist uniquely $\tilde{E}(x,\lambda)$ and $\tilde{g}(x,\lambda)$ so that

$$g(\lambda)E(x,\lambda) = \tilde{E}(x,\lambda)\tilde{g}(x,\lambda), \tag{1.5}$$

$\tilde{E}(x,\lambda)$ is holomorphic for $\lambda \in C$, $\tilde{g}(x,\lambda)$ is holomorphic near $\lambda = \infty$ and $\tilde{g}(x,\infty) = I$. Calculate the residue at $\lambda = \infty$ to conclude that

$$\tilde{E}^{-1}d\tilde{E} = \sum_{i=1}^{n}(a_i\lambda + [a_i, \tilde{v}])dx_i$$

for some $\tilde{v}$. So $\tilde{v}$ is a new solution of (1.1). The solution $\tilde{v}$ can also be obtained from $\tilde{g}$ as follows: Expand

$$\tilde{g}(x,\lambda) \;=\; I + m_1(x)\lambda^{-1} + m_2(x)\lambda^{-2} + \cdots$$

at $\lambda = \infty$. Then

$$\tilde{v} = v - p_0(m_1),$$

where p_0 is the projection onto $\mathcal{A}^\perp \cap \mathcal{P}$. The map

$$g\sharp v = \tilde{v}$$

defines the dressing action of the group of germs of holomorphic maps on the space of local solutions of the U/K-system.

If $g(\lambda)$ is a meromorphic map on S^2 with $g(\infty) = I$, then the factorization (1.5) can be done explicitly by calculating the residues at poles of $g(\lambda)$. Note that system (1.1) has a trivial solution $v = 0$ and

$$E(x,\lambda) = \exp\left(\sum_{i=1}^{n} a_i x_i \lambda\right)$$

is the solution of the corresponding linear system (1.4). Therefore $g\sharp 0$ can be computed explicitly. These explicit solutions correspond to the "*pure solitons*" in the theory of soliton equations.

If $g(\lambda)$ is a holomorphic map defined in a neighborhood of ∞ in S^2 such that $g(\lambda)a_1g(\lambda)^{-1}$ is a polynomial in λ^{-1}, then the solution $g\sharp 0$ can be obtained by solving a system of ordinary differential equations on a finite dimensional linear space. These solutions are the so called "*finite type solutions*". Finite type solutions have been used successfully to construct constant mean curvature tori in R^3 by Pinkall and Sterling in [PiS], in 3-dimensional space forms by Bobenko in [Bo1], and harmonic maps from a torus to a symmetric space by Burstall, Ferus, Pedit and Pinkall in [BFPP].

- **Cauchy problem**

If a_1 is regular, then the linear map $\mathrm{ad}(a_1) : \mathcal{P}\cap\mathcal{A}^\perp \to \mathcal{K}$ is injective. It follows from Cartan-Kähler Theorem that if $v_0 : (-\delta, \delta) \to \mathcal{P}\cap\mathcal{A}^\perp$ is real analytic, then system (1.1) has a unique local analytic solution such that $v(x_1, 0, \cdots, 0) = v_0(x_1)$. If the initial data v_0 is not real analytic but is rapidly decaying, then we can use the inverse scattering method developed by Beals and Coifman [BC] to solve the initial value problem (cf. [TU1], [Te2]).

- **Gauge equivalent systems**

Let θ_λ be the flat connection (1.3) associated to the solution v of the U/K-system (1.1), and $g : R^n \to U_C$ a smooth map. Then the gauge transformation

$$g * \theta_\lambda = g\theta_\lambda g^{-1} - dgg^{-1}$$

is again a flat connection 1-form for all $\lambda \in C$. However, the differential equation given by the condition that $g * \theta_\lambda$ is flat for all λ has a different form. We say this new equation is gauge equivalent to system (1.1). For example:

(i) Since θ_λ satisfies the U/K-reality condition, θ_0 is a $\mathcal{K}$-valued flat connection 1-form. Hence there exists g such that $g^{-1}dg = \theta_0$. A direct computation shows that the gauge transformation of θ_λ by g is

$$g * \theta_\lambda = g\theta_\lambda g^{-1} - dgg^{-1} = \sum_{i=1}^{n} ga_ig^{-1}\lambda dx_i.$$

Write $A_i = ga_ig^{-1}$. The equation given by the flatness of $g * \theta_\lambda$ is the *curved flat system* studied by Ferus and Pedit in [FP1].

(ii) Suppose $K = K_1 \times K_2$. Let v be a solution of the U/K-system, and $g^{-1}dg = \theta_0$. Since $\theta_0 \in \mathcal{K}$, $g(x) \in K = K_1 \times K_2$. So we can write $g = (g_1, g_2) \in K_1 \times K_2$. A direct computation shows that the coefficients of λ^0 in $g_1 * \theta_\lambda$ and $g_2 * \theta_\lambda$ are in $\mathcal{K}_2$ and $\mathcal{K}_1$ respectively. The equations given by the flatness of $g_1 * \theta_\lambda$ and $g_2 * \theta_\lambda$ are called the U/K-system I and II respectively.

- **Gauss-Codazzi equations for Submanifolds in space forms**

Let $O(n, 1)$ denote the group of all $g \in GL(n+1)$ that preserves the bilinear form

$$x_1^2 + \cdots + x_n^2 - x_{n+1}^2.$$

Henceforth in this paper, we use the following notations:

$$G_{m,n} = O(m+n)/O(m) \times O(n), \quad G^1_{m,n} = O(m+n, 1)/O(m) \times O(n, 1).$$

Let $N^n(c)$ denote the n-dimensional space form with curvature c, i.e., the complete, simply connected Riemannian manifold with constant sectional curvature c. So $N^n(c)$ is R^n, S^n and H^n for $c = 0, 1, -1$ respectively. The Levi-Civita connection 1-form of $N^n(c)$ can be read from the flatness of a $so(n)$, $so(n+1)$ and $o(n,1)$- valued connection 1-form. The Gauss-Codazzi equation of a submanifold in $N^n(c)$ is given by the flatness of the restriction of this connection 1-form to the submanifold. The Fundamental Theorem of Submanifolds states that each solution of the Gauss-Codazzi equation correspond to a submanifold in $N^n(c)$, unique up to ambient isometry. So if v is a solution of the $G_{m,n}$- or $G^1_{m,n}$-system I or II, then the corresponding Lax connection θ_λ at $\lambda = 1$ gives rise to a submanifold of a certain space form. Using the method of moving frames, special properties of the flat connection θ_1 can be translated easily to geometric properties of the corresponding submanifolds.

- **Submanifolds corresponding to the $G_{m,n}$- and $G^1_{m,n}$-system I**

In [Te2], Terng proved that solutions of the $G_{n,n}$-, $G_{n,n+1}$- and $G^1_{n,n}$-system I correspond to local isometric immersions of the space form $N^n(c)$ in $N^{2n}(c)$ with flat normal bundle for $c = 0, 1$, and -1 respectively. We generalize this result to the $G_{m,n}$- and $G^1_{m,n}$-system I for any $m \geq n$. They give rise to local isometric immersions of $N^n(c)$ into $N^{n+m}(c)$.

- **Submanifolds corresponding to $G_{m,n}$- and $G^1_{m,n}$-system II**

In order to explain the submanifold geometry corresponding to the $G_{m,n}$- and $G^1_{m,n}$-system II, we first need to review some classical surface theory. Let M be a surface in R^3 with curvature $K = -1$, and e_3 its unit normal field. Then there exists a line of curvature coordinate system (x, y) such that the two fundamental forms of M are

$$I_1 = \cos^2 u \; dx^2 + \sin^2 u \; dy^2, \quad II_1 = \sin u \cos u \; (dx^2 - dy^2).$$

Since the unit sphere is totally umbilic, the fundamental forms for S^2 in (x, y) coordinates via the parametrization $e_3(x, y)$ are

$$I_2 = \sin^2 u \; dx^2 + \cos^2 u \; dy^2, \quad II_2 = -(\sin^2 u \; dx^2 + \cos^2 u \; dy^2).$$

The Gauss-Codazzi equations for M (curvature -1) and e_3 (curvature 1) are the same sine-Gordon equation

$$u_{xx} - u_{yy} = \sin u \cos u, \tag{SGE}$$

and the tangent plane of M at (x, y) is the same as the tangent plane of the sphere at $e_3(x, y)$. A direct computation shows that such u gives rise to a solution of the $G_{3,2}$-system II. This is a special case. In fact, we show that each solution of the $G_{3,2}$-system II corresponds to a pair of surfaces (X_1, X_2) in R^3 with common line of curvature coordinates x, y such that the tangent plane at $X_1(x, y)$ is equal to the tangent plane at $X_2(x, y)$ and their fundamental forms are

$$\begin{aligned} I_1 &= \cos^2 u \; dx^2 + \sin^2 u \; dy^2, \quad II_1 = g_1 \cos u \; dx^2 + g_2 \sin u \; dy^2, \\ I_2 &= \sin^2 u \; dx^2 + \cos^2 u \; dy^2, \quad II_2 = -g_1 \sin u \; dx^2 + g_2 \cos u \; dy^2 \end{aligned}$$

for some functions u, g_1, g_2. Moreover, the Gaussian curvature

$$K_1(x,y) = -K_2(x,y).$$

For general $m > n$, we prove that each solution of the $G_{m,n}$-system II ($G^1_{m,n-1}$-system II respectively) gives rise to an n-tuple $(X_1, \cdots, X_n)$ of n-dimensional submanifolds in R^m with flat normal bundles and common line of curvature coordinates $(x_1, \cdots, x_n)$ such that fundamental forms of X_j are

$$I_j = \sum_{i=1}^{n} a_{ji}^2(x) dx_i^2, \quad II_j = \sum_{i=1,k=1}^{n,m-n} a_{ji} g_{ki}\, dx_i^2 e_{n+k}$$

for some $(a_{ij}(x)) \in O(n)$ ($\in O(n-1,1)$ respectively) and $g_{ki}(x)$.

If we use another standard form of $O(n+1,1)$, the group of $g \in GL(n+2)$ that leaves the bi-linear from

$$x_1^2 + \cdots + x_n^2 + 2x_{n+1}x_{n+2}$$

invariant, then the corresponding 2-tuple (X_1, X_2) in R^n for the $G^1_{n,1}$-system II has the property that X_1 is an isothermic surface and X_2 is a Christoffel dual of X_1. This was proved by Burstall, Hertrich-Jeromin, Pedit and Pinkall in [BHPP] for $n = 2$ and by Burstall in [Bu] for general n.

- **Bäcklund transformations and dressing action**

Let M, M^* be two surfaces in R^3. A diffeomorphism $\ell : M \to M^*$ is called a *Bäcklund transformation* with constant θ if for all $p \in M$,
(a) $\overline{pp^*}$ is tangent to both M and M^* at p and $p^* = \ell(p)$,
(b) $\mathrm{d}(p, p^*) = \sin\theta$,
(c) the angle between TM_p and $TM^*_{p^*}$ is θ.
Bäcklund proved ([Ba]) that if ℓ is a Bäcklund transformation, then both M and M^* have curvature -1. Moreover, if M is a surface in R^3 with $K = -1$, $0 < \theta < \pi$ a constant, and $v_0 \in TM_{p_0}$ a unit vector that is not a principal direction, then there exist a unique surface M^* and a Bäcklund transformation $\ell : M \to M^*$ such that $\ell(p_0) = p_0 + \sin\theta\ v_0$. Analytically, this gives a method of constructing new solution of SGE from a given one. Bäcklund transformations have been generalized to isometric immersions of $N^n(c)$ in $N^{2n-1}(c+1)$ by Terng and Tenenblat for $c = -1$ in [TT] and by Tenenblat for $c = 0, 1$ in [Ten].

Terng and Uhlenbeck proved in [TU2] that the dressing action of a meromorphic map with one pole on the space of solutions of SGE gives rise exactly to the classical Bäcklund transformations. We generalize this result to $G_{n,n}$ and $G^1_{n,n}$-systems.

- **Ribaucour transformations and dressing action**

Let $M, \tilde{M}$ be two surfaces in R^3. A diffeomorphism $\ell : M \to \tilde{M}$ is called a *Ribaucour transformation* if for all $p \in M$
(a) the normal line at p to M meets the normal line at $\ell(p)$ to $\tilde{M}$ at equidistance $r(p)$,

(b) if e is a principal direction of M at p, then $\ell_*(e)$ is parallel to a principal direction of $\tilde{M}$ at $\ell(p)$ and the line through p in the direction e meets the line through $\ell(p)$ in the direction of $\ell_*(e)$ at a point at equidistance $s(p)$.

The notion of Ribaucour transformations has a natural generalization to submanifolds in space forms with flat normal bundle ([DT]). We show that the dressing action of rational maps with two simple poles on the solutions of the $G_{m,n}$- and $G^1_{m,n}$-system I and II correspond to Ribaucour transformations for submanifolds.

• **Organization of the paper**

We review some general facts about the U/K-system in section 2, write down the $G_{m,n}$-systems explicitly in section 3, and the $G^1_{m,n}$-systems in section 4. We review the method of moving frames in section 5. We describe submanifolds associated to various $G_{m,n}$-systems and $G^1_{m,n}$-systems in section 6 and 7 respectively. In section 8, we study relations between isothermic surfaces and $G^1_{m,1}$-systems. The dressing action of a rational map with two simple poles on solutions of the $G_{m,n}$- and $G^1_{m,n}$-systems are written down explicitly in section 9 and 11 respectively. The corresponding geometric transformations are given in section 10 and 12. Burstall [Bu] gave a generalization of isothermic surfaces in R^3 and their Darboux transformations to isothermic surfaces in R^n. In section 13, we show that the dressing action of a rational map with two poles on the space of solutions of the $G^1_{n,1}$-system II gives rise to these Darboux transformations. In section 14, we give a relation between the dressing action of loop with one simple pole and Bäcklund transformations. A permutability formula for Ribaucour transformations is explained in section 15, and relation between dressing action and finite type solutions of the U/K-system are given in the last section.

2. The U/K-System

A connection of a trivial principal U-bundle over R^n is:

$$\frac{\partial}{\partial x_i} + A_i, \quad 1 \le i \le n,$$

for some smooth maps $A_i : R^n \to \mathcal{U}$. The curvature of this connection is

$$\Omega_{ij} = \left[\frac{\partial}{\partial x_i} + A_i, \frac{\partial}{\partial x_j} + A_j\right] = (A_j)_{x_i} - (A_i)_{x_j} + [A_i, A_j].$$

A connection is flat if its curvature is zero.

The following Proposition, which is well-known, gives several equivalent conditions for a connection to be flat. The proof follows from a direct computation.

2.1 Proposition. *Let $A_1, \cdots, A_n : R^n \to \mathcal{U}$ be smooth maps. The following statements are equivalent:*

(1) the connection $\frac{\partial}{\partial x_i} + A_i(x)$ is flat, i.e., $[\frac{\partial}{\partial x_i} + A_i, \frac{\partial}{\partial x_j} + A_j] = 0$,

(2) $(A_j)_{x_i} - (A_i)_{x_j} + [A_i, A_j] = 0$,

(3) *the connection 1-form* $\theta = \sum_{i=1}^{n} A_i dx_i$ *is flat, i.e.,* $d\theta + \theta \wedge \theta = 0$.

(4)

$$E_{x_i} = EA_i, \quad 1 \le i \le n, \tag{2.1}$$

is solvable for $E : R^n \to U$.

E is called a *trivialization* of the flat connection $\theta = \sum_i A_i dx_i$ if it is a solution of (2.1) or equivalently if

$$E^{-1}dE = \sum_{i=1}^{n} A_i dx_i.$$

It follows from a direct computation and Proposition 2.1 that

2.2 Proposition ([Te2]). *The following statements are equivalent for a map* $v : R^n \to \mathcal{P} \cap \mathcal{A}^{\perp}$:

(i) v *is a solution of the* U/K*-system (1.1),*

(ii)

$$\left[\frac{\partial}{\partial x_i} + \lambda a_i + [a_i, v], \frac{\partial}{\partial x_j} + \lambda a_j + [a_j, v]\right] = 0, \quad \forall\, \lambda \in C, \tag{2.2}$$

(iii) θ_λ *is a flat* $\mathcal{U}_C = \mathcal{U} \otimes C$*-connection 1-form on* R^n *for all* $\lambda \in C$*, where*

$$\theta_\lambda = \sum_{i=1}^{n} (a_i \lambda + [a_i, v]) dx_i, \tag{2.3}$$

(iv) *there exists* E *so that* $E^{-1}dE = \theta_\lambda$.

The one parameter family of flat connections (2.2) or (2.3) is called the *Lax connection* of the U/K-system (1.1).

An element $b \in \mathcal{P}$ is called *regular* if the orbit at b for the $\mathrm{Ad}(K)$-action is a principal orbit. Let $\mathcal{A}$ be a maximal abelian subalgebra in $\mathcal{P}$,

$$\mathcal{K}_{\mathcal{A}} = \{\xi \in \mathcal{K} \mid [\xi, a] = 0 \quad \forall\, a \in \mathcal{A}\},$$

and $\mathcal{K}_{\mathcal{A}}^{\perp}$ the orthogonal complement of $\mathcal{K}_{\mathcal{A}}$ in $\mathcal{K}$. It follows from standard theory of symmetric space (cf. [H]) that if $b \in \mathcal{A}$ is regular, then $\mathrm{ad}(b)$ maps $\mathcal{K}_{\mathcal{A}}^{\perp}$ and $\mathcal{P} \cap \mathcal{A}^{\perp}$ isomorphically to $\mathcal{P} \cap \mathcal{A}^{\perp}$ and $\mathcal{K}_{\mathcal{A}}^{\perp}$ respectively.

2.3 Proposition. *Let* $a_1, \cdots, a_n$ *be a basis of a maximal abelian subalgebra* $\mathcal{A}$ *in* $\mathcal{P}$, *and* $u_i : R^n \to \mathcal{K}_{\mathcal{A}}^{\perp}$ *smooth maps for* $1 \le i \le n$. *If*

$$\theta_\lambda = \sum_{i=1}^{n} (a_i \lambda + u_i) dx_i \tag{2.4}$$

is a flat connection 1-form on R^n *for all* $\lambda \in C$, *then there exists a unique map* $v : R^n \to \mathcal{P} \cap \mathcal{A}^{\perp}$ *such that* $u_i = [a_i, v]$.

PROOF. Choose a basis $b_1, \cdots, b_n$ of $\mathcal{A}$ such that each b_i is regular. Write $b_j = \sum_i c_{ij} a_i$. Make a change of coordinates $x_i = \sum_j c_{ij} y_j$. Then

$$\theta_\lambda = \sum_i (a_i \lambda + u_i) dx_i = \sum_j (b_j \lambda + \tilde{u}_j) dy_j,$$

where $\tilde{u}_j = \sum_i c_{ij} u_i \in \mathcal{K} \cap \mathcal{A}^\perp$. Note that θ_λ is flat for all λ if and only if

$$\begin{cases} [b_i, \tilde{u}_j] = [b_j, \tilde{u}_i], \\ (\tilde{u}_j)_{x_i} - (\tilde{u}_i)_{x_j} + [\tilde{u}_i, \tilde{u}_j] = 0. \end{cases} \tag{2.5}$$

Because $b_1, \cdots, b_n$ are regular, $\mathrm{ad}(b_j)$ maps $\mathcal{P} \cap \mathcal{A}^\perp$ isomorphically to $\mathcal{K} \cap \mathcal{A}^\perp$. Hence there exists a unique $v_j \in \mathcal{P} \cap \mathcal{A}^\perp$ such that $\tilde{u}_j = \mathrm{ad}(b_j)(v_j)$ for $1 \leq j \leq n$. Then the first equation of (2.5) implies that

$$\mathrm{ad}(b_i)\,\mathrm{ad}(b_j)(v_j) = \mathrm{ad}(b_j)\,\mathrm{ad}(b_i)(v_i).$$

Since $[b_i, b_j] = 0$, $\mathrm{ad}(b_i)\,\mathrm{ad}(b_j) = \mathrm{ad}(b_j)\,\mathrm{ad}(b_i)$. But $\mathrm{ad}(b_i)$ is injective on $\mathcal{P} \cap \mathcal{A}^\perp$ implies that $v_i = v_j$, which will be denoted by v. We compute directly to get

$$\begin{aligned} u_j &= \sum_i c^{ij} \tilde{u}_i = \sum_i c^{ij} [b_i, v] \\ &= \sum_i c^{ij} \left[\sum_k c_{ki} a_k, v \right] = [a_j, v], \end{aligned}$$

where (c^{ij}) is the inverse of (c_{ij}). ∎

The following Proposition is immediate:

2.4 Proposition. *If θ is a flat $\mathcal{G}$-valued connection 1-form and $g : R^n \to G$ a smooth map, then the gauge transformation of θ by g,*

$$g * \theta = g\theta g^{-1} - dg g^{-1},$$

is also flat. Moreover, if E is a trivialization of θ, then Eg^{-1} is a trivialization of $g * \theta$.

Consider the system of PDE for

$$(A_1, \cdots, A_n, B_1, \cdots, B_n) : R^n \to \prod_{i=1}^n \mathcal{P} \times \prod_{i=1}^n \mathcal{K}$$

given by the condition that

$$\sum_{i=1}^n (A_i \lambda + B_i) dx_i$$

is a flat connection on R^n for all $\lambda \in C$. Or equivalently,

$$(A_j\lambda + B_j)_{x_i} - (A_i\lambda + B_i)_{x_j} + [A_i\lambda + B_i, A_j\lambda + B_j] = 0$$

for all $\lambda \in C$. By comparing coefficients of λ^2, λ and the constant term, we get

$$\begin{cases} [A_i, A_j] = 0, \\ (A_i)_{x_j} - (A_j)_{x_i} = [A_i, B_j] + [B_i, A_j], \\ (B_i)_{x_j} - (B_j)_{x_i} = [B_i, B_j]. \end{cases} \tag{2.6}$$

2.5 Proposition. *Let $\Omega_\lambda = \sum_i (A_i\lambda + B_i)dx_i$, $A_i \in \mathcal{P}$ and $B_i \in \mathcal{K}$. If $[A_i, A_j] = 0$ for all $1 \leq i, j \leq n$, then Ω_λ is flat for all $\lambda \in C$ if and only if Ω_{λ_0} is flat for some non-zero real or pure imaginary λ_0.*

PROOF. If Ω_{λ_0} is flat, then

$$(\lambda_0 A_j + B_j)_{x_i} - (\lambda_0 A_i + B_i)_{x_j} + [\lambda_0 A_i + B_i, \lambda_0 A_j + B_j] = 0. \tag{2.7}$$

So both the $\mathcal{K}$ and $\mathcal{P}$ components of the left hand side of the equation must be zero. Since U/K is a symmetric space,

$$[\mathcal{K}, \mathcal{K}] \subset \mathcal{K}, \quad [\mathcal{K}, \mathcal{P}] \subset \mathcal{P}, \quad [\mathcal{P}, \mathcal{P}] \subset \mathcal{K}.$$

Equate the $\mathcal{K}$ and $\mathcal{P}$ components of (2.7) to get

$$\begin{cases} (B_j)_{x_i} - (B_i)_{x_j} + [B_i, B_j] = 0, \\ (A_j)_{x_i} - (A_i)_{x_j} + [A_i, B_j] + [B_i, A_j] = 0. \end{cases}$$

But this is the equation for Ω_λ to be flat for all $\lambda \in C$. ∎

Restrict system (2.6) to the case when all $B_i = 0$ to get a system for maps $(A_1, \cdots, A_n) : R^n \to \mathcal{P} \times \cdots \times \mathcal{P}$:

$$\begin{cases} [A_i, A_j] = 0, \\ (A_i)_{x_j} = (A_j)_{x_i}, \quad \text{for all } i \neq j. \end{cases} \tag{2.8}$$

This is the *Curved Flat system* associated to U/K defined by Ferus and Pedit in [FP]. Its Lax connection is

$$\left[\frac{\partial}{\partial x_i} + A_i\lambda, \frac{\partial}{\partial x_j} + A_j\lambda\right] = 0, \quad \forall\, \lambda \in C.$$

The second equation of (2.8) implies that $\sum_i A_i dx_i$ is exact. So we get

2.6 Proposition. *Let $A_i(x) \in \mathcal{P}$. Then $\sum_i \lambda A_i(x) dx_i$ is flat for all $\lambda \in C$ if and only if $[A_i, A_j] = 0$ for all $1 \leq i, j \leq n$ and there exists a map $X : R^n \to \mathcal{P}$ such that*

$$dX = \sum_{i=1}^n A_i dx_i.$$

Let θ_λ be defined as in (2.4). If θ_λ is flat for all $\lambda \in C$, then $\theta_0 = \sum_i u_i dx_i$ is flat. Let g be a trivialization of θ_0. A direct computation shows that the gauge transformation of θ_λ by g is

$$g * \theta_\lambda = \sum_{i=1}^{n} g a_i g^{-1} \lambda dx_i.$$

In other words, we have gauged away the $\mathcal{K}$-part of the Lax connection θ_λ and the corresponding PDE is the curved flat system. If $\mathcal{K} = \mathcal{K}_1 \oplus \mathcal{K}_2$, then we can gauge away the $\mathcal{K}_1$-part ($\mathcal{K}_2$-part respectively) of θ_λ. To do this, we write

$$u_i = \xi_i + \eta_i \in \mathcal{K}_1 + \mathcal{K}_2.$$

Since $\theta_0 = \sum_i u_i dx_i$ is flat, both $\sum_i \xi_i dx_i$ and $\sum_i \eta_i dx_i$ are flat. Let $g_1 : R^n \to K_1$ and $g_2 : R^n \to K_2$ be trivializations of $\sum_i \xi_i dx_i$ and $\sum_i \eta_i dx_i$ respectively, i.e.,

$$g_1^{-1} dg_1 = \sum_i \xi_i dx_i, \quad g_2^{-1} dg_2 = \sum_i \eta_i dx_i.$$

Then the gauge transformation of θ_λ by g_1 and g_2 are

$$g_1 * \theta_\lambda = \sum_i (g_1 a_i g_1^{-1} \lambda + \eta_i) dx_i,$$
$$g_2 * \theta_\lambda = \sum_i (g_2 a_i g_2^{-1} \lambda + \xi_i) dx_i,$$

respectively.

The $U/(K_1 \times K_2)$-*system* I is the PDE for $g_1 : R^n \to K_1$ and $\eta_1, \cdots, \eta_n : R^n \to \mathcal{K}_2 \cap \mathcal{K}_{\mathcal{A}}^\perp$ such that

$$\theta_\lambda^I = \sum_i (g_1 a_i g_1^{-1} \lambda + \eta_i) dx_i$$

is flat for all $\lambda \in C$, i.e.,

$$\begin{cases} [g_1^{-1}(g_1)_{x_i}, a_j] - [g_1^{-1}(g_1)_{x_j}, a_i] + [a_i, g_1^{-1}\eta_j g_1] + [g_1^{-1}\eta_i g_1, a_j] = 0, \\ (\eta_j)_{x_i} - (\eta_i)_{x_j} + [\eta_i, \eta_j] = 0. \end{cases} \tag{2.9}$$

Similarly, the $U/(K_1 \times K_2)$-*system* II is the PDE for $g_2 : R^n \to K_2$ and $\xi_1, \cdots, \xi_n : R^n \to \mathcal{K}_1 \cap \mathcal{K}_{\mathcal{A}}^\perp$ such that

$$\theta_\lambda^{II} = \sum_i (g_2 a_i g_2^{-1} \lambda + \xi_i) dx_i$$

is flat for all $\lambda \in C$, i.e.,

$$\begin{cases} [g_2^{-1}(g_2)_{x_i}, a_j] - [g_2^{-1}(g_2)_{x_j}, a_i] + [a_i, g_2^{-1}\xi_j g_2] + [g_2^{-1}\xi_i g_2, a_j] = 0, \\ (\xi_j)_{x_i} - (\xi_i)_{x_j} + [\xi_i, \xi_j] = 0. \end{cases} \tag{2.10}$$

3. $G_{m,n}$-systems

In this section, we assume

$$U/K = G_{m,n} = O(m+n)/(O(m)\times O(n)) \quad \text{with } m \geq n.$$

We write down the $G_{m,n}$-systems I and II explicitly.

Let $\mathcal{U} = o(m+n)$, and $\sigma : \mathcal{U} \to \mathcal{U}$ be the involution defined by $\sigma(X) = I_{m,n}^{-1} X I_{m,n}$, where

$$I_{m,n} = \begin{pmatrix} I_m & 0 \\ 0 & -I_n \end{pmatrix}.$$

Let $\mathcal{K}$ and $\mathcal{P}$ denote the $+1$ and -1 eigenspaces of σ respectively. Then $\mathcal{U} = \mathcal{K} + \mathcal{P}$ is a Cartan decomposition of U/K, where

$$\mathcal{K} = o(m)\times o(n) = \left\{ \begin{pmatrix} Y_1 & 0 \\ 0 & Y_2 \end{pmatrix} \middle| \, Y_1 \in o(m),\, Y_2 \in o(n) \right\},$$

$$\mathcal{P} = \left\{ \begin{pmatrix} 0 & \xi \\ -\xi^t & 0 \end{pmatrix} \middle| \, \xi \in \mathcal{M}_{m\times n} \right\}.$$

Here $\mathcal{M}_{m\times n}$ is the set of $m\times n$ matrices. Note that

$$\mathcal{A} = \left\{ \begin{pmatrix} 0 & -D \\ D^t & 0 \end{pmatrix} \middle| \, D = (d_{ij}) \in \mathcal{M}_{m\times n},\, d_{ij} = 0 \text{ if } i \neq j \right\}.$$

is a maximal abelian subalgebra in $\mathcal{P}$ and

$$\mathcal{P}\cap\mathcal{A}^{\perp} = \left\{ \begin{pmatrix} 0 & \xi \\ -\xi^t & 0 \end{pmatrix} \middle| \, \xi = (\xi_{ij}) \in \mathcal{M}_{m\times n},\, \xi_{ii} = 0 \text{ for } 1 \leq i \leq n \right\}.$$

Let

$$a_i = \begin{pmatrix} 0 & -D_i \\ D_i^t & 0 \end{pmatrix},$$

where $D_i \in \mathcal{M}_{m\times n}$ is the matrix all whose entries are zero except the (i,i)-th entry is equal to 1. Then $a_1, \ldots, a_n$ form a basis of $\mathcal{A}$. The U/K-system (1.1) for this symmetric space is the following PDE for $\xi = (\xi_{ij}) : R^n \to \mathcal{M}_{m\times n}$ with $\xi_{ii} = 0$ for all $1 \leq i \leq n$:

$$\begin{cases} D_i\xi^t_{x_j} - \xi_{x_j}D_i^t - D_j\xi^t_{x_i} + \xi_{x_i}D_j^t = [D_i\xi^t - \xi D_i^t, D_j\xi^t - \xi D_j^t], & i \neq j, \\ D_i^t\xi_{x_j} - \xi^t_{x_j}D_i - D_j^t\xi_{x_i} + \xi^t_{x_i}D_j = [D_i^t\xi - \xi^t D_i, D_j^t\xi - \xi^t D_j], & i \neq j. \end{cases} \tag{3.1}$$

Its Lax connection (2.3) is

$$\theta_\lambda = \sum_i \left\{ \lambda \begin{pmatrix} 0 & -D_i \\ D_i^t & 0 \end{pmatrix} + \begin{pmatrix} D_i\xi^t - \xi D_i^t & 0 \\ 0 & -\xi^t D_i + D_i^t \xi \end{pmatrix} \right\} dx_i. \tag{3.2}$$

Next we write down the $G_{m,n}$-system I and II explicitly. Let

$$g = \begin{pmatrix} A & 0 \\ 0 & B \end{pmatrix} \in O(m) \times O(n)$$

be a solution of

$$g^{-1}dg = \theta_0 = \sum_{i=1}^{n} \begin{pmatrix} D_i\xi^t - \xi D_i^t & 0 \\ 0 & -\xi^t D_i + D_i^t \xi \end{pmatrix} dx_i.$$

Let

$$g_1 = \begin{pmatrix} A & 0 \\ 0 & I \end{pmatrix}, \quad g_2 = \begin{pmatrix} I & 0 \\ 0 & B \end{pmatrix}.$$

Write

$$\xi = \begin{pmatrix} F \\ G \end{pmatrix}, \quad D_i = \begin{pmatrix} C_i \\ 0 \end{pmatrix}, \quad A = (A_1, A_2),$$

where $F, C_i \in gl(n)$, $G \in \mathcal{M}_{(m-n)\times n}$, $A_1 \in \mathcal{M}_{m\times n}$, and $A_2 \in \mathcal{M}_{m\times(m-n)}$. Then

$$g_1 a_i g_1^{-1} = \begin{pmatrix} 0 & -A_1 C_i \\ C_i A_1^t & 0 \end{pmatrix}.$$

Let

$$\begin{aligned} \mathcal{M}^0_{m\times n} &= \{A_1 \in \mathcal{M}_{m\times n} \mid A_1^t A_1 = I\}, \\ gl(n)_* &= \{(x_{ij}) \in gl(n) \mid x_{ii} = 0, \ 1 \le i \le n\}. \end{aligned}$$

The *U/K-system I* is the PDE for $(A_1, F) : R^n \to \mathcal{M}^0_{m\times n} \times gl_*(n)$ such that

$$\theta^I_\lambda = \sum_{i=1}^{n} \begin{pmatrix} 0 & -A_1 C_i \lambda \\ C_i A_1^t \lambda & -F^t C_i + C_i^t F \end{pmatrix} dx_i \tag{3.3}$$

is flat for all $\lambda \in C$, i.e.,

$$\begin{cases} (a_{ij})_{x_k} = f_{jk} a_{ik}, & \text{if } k \neq j, \\ (f_{ij})_{x_j} + (f_{ji})_{x_i} + \sum_k f_{ik} f_{jk} = 0, & \text{if } i \neq j, \\ (f_{ij})_{x_k} = f_{ik} f_{kj}, & \text{if } i, j, k \text{ are distinct}, \end{cases} \tag{3.4}$$

where $A = (a_{ij})$ and $F = (f_{ij})$. Note that equation (3.4) is the condition that the above θ^I_λ is flat for $\lambda = 1$. So we have

3.1 Proposition. *The following statements are equivalent for map (A_1, F) : $R^n \to \mathcal{M}^0_{m\times n} \times gl_*(n)$:*
(i) (A_1, F) is a solution of the $G_{m,n}$-system I (3.4).
(ii) θ^I_λ defined by (3.3) is flat for all $\lambda \in C$.
(iii) θ^I_λ defined by (3.3) is flat for $\lambda = 1$.

Note that if $a_{ij} \neq 0$ for all $1 \leq j \leq n$, then the first set of equations of (3.4) implies that F can be computed from the i-th row of A_1 by $f_{jk} = (a_{ij})_{x_k}/a_{ik}$.

Next we explain the reality conditions. Recall that a symmetric space U/K is determined by a conjugate linear Lie algebra involution τ and a complex linear Lie algebra involution σ on the complexified Lie algebra $\mathcal{U}_{\boldsymbol{C}} = \mathcal{U} \otimes \boldsymbol{C}$ such that

(i) τ and σ commute,

(ii) $\mathcal{U}$ is the fixed point set of τ, and $\mathcal{K}$ and $\mathcal{P}$ are the $+1, -1$ eigenspaces of σ on $\mathcal{U}$ respectively, and $\mathcal{U} = \mathcal{K} + \mathcal{P}$ is the Cartan decomposition.

We still use τ and σ to denote the corresponding involutions on the group $U_{\boldsymbol{C}}$.

A map $g : C \to U_{\boldsymbol{C}}$ ($g : C \to \mathcal{U}_{\boldsymbol{C}}$ respectively) is said to satisfy the *U/K-reality condition* if

$$\tau(g(\bar{\lambda})) = g(\lambda), \quad \sigma(g(-\lambda)) = g(\lambda). \tag{3.5}$$

A direct computation gives

3.2 Proposition.

(i) If $A(\lambda) = \sum_i A_i \lambda^i : C \to \mathcal{U}_C$ satisfies the U/K-reality condition, then $A_i \in \mathcal{K}$ if i is even and is in $\mathcal{P}$ if i is odd.

(ii) The Lax pair θ_λ defined by (2.3) for the U/K-system (1.1) satisfies the U/K-reality condition.

3.3 Definition. A *frame* for a solution v of the U/K-system (I, II respectively) is a trivialization of the corresponding Lax connection θ_λ ($\theta^I_\lambda, \theta^{II}_\lambda$ respectively) that satisfies the U/K-reality condition.

3.4 Remark.

(i) If $g : C \to U_C$ satisfies the U/K-reality condition, then $g(0) \in K$.

(ii) If $E(x, \lambda)$ is the trivialization of θ_λ defined by (2.3) such that $E(0, \lambda)$ satisfies the U/K-reality condition, then E also satisfies the U/K-reality condition.

Let $p \in O(m)$ be a constant matrix. The gauge transform of θ^I_λ by $g = \begin{pmatrix} p & 0 \\ 0 & I \end{pmatrix}$ is

$$g * \theta^I_\lambda = \begin{pmatrix} 0 & -pA_1 C_i \lambda \\ C_i^t A_1^t p^t \lambda & -F^t C_i + C_i^t F \end{pmatrix},$$

which is flat for all $\lambda \in C$. Note that the coefficient of λ in $g * \theta_\lambda$ lies in $\mathcal{P}$, the constant term lies in $\mathcal{K}$, and

$$(pA_1)^t (pA_1) = A_1^t p^t p A_1 = A_1^t A_1 = I.$$

So it follows from Proposition 3.1 that

3.5 Corollary. *Let (A_1, F) be a solution of the $G_{m,n}$-system I (3.4), and $p \in O(m)$ a constant matrix. Then (pA_1, F) is also a solution of (3.4). Moreover, if E^I is a frame for (A_1, F), then $E^I p^{-1}$ is a frame for (pA_1, F).*

3.6 Proposition.

(i) Suppose $\xi = \begin{pmatrix} F \\ G \end{pmatrix}$ *is a solution of the* $G_{m,n}$*-system (3.1),* θ_λ *the corresponding Lax connection, and* $g = \begin{pmatrix} A & 0 \\ 0 & B \end{pmatrix} : R^n \to O(m) \times O(n)$ *satisfies* $g^{-1}dg = \theta_0$. *Write* $A = (A_1, A_2)$ *with* $A_1 \in \mathcal{M}_{m\times n}$. *Then* (A_1, F) *is a solution of the* $G_{m,n}$*-system I (3.4).*

(ii) Conversely, if (A_1, F) *is a solution of the* $G_{m,n}$*-system I (3.4), then there exists an* $\mathcal{M}_{(m-n)\times n}$*-valued map* G *such that* $\xi = \begin{pmatrix} F \\ G \end{pmatrix}$ *is a solution of (3.1).*

PROOF.

(i) follows from the definition of the U/K-system I.

(ii) Choose A_2 so that $A = (A_1, A_2) \in O(m)$. Let $g = \begin{pmatrix} A^t & 0 \\ 0 & I \end{pmatrix}$. Then the gauge transformation of g on θ^I_λ is

$$g{*}\theta^I_\lambda = g\theta^I_\lambda g^{-1} - dgg^{-1}$$
$$= \sum_i \left\{ \lambda \begin{pmatrix} 0 & 0 & -C_i \\ 0 & 0 & 0 \\ C_i & 0 & 0 \end{pmatrix} + \begin{pmatrix} A_1^t\,(A_1)_{x_i} & A_1^t\,(A_2)_{x_i} & 0 \\ A_2^t\,(A_1)_{x_i} & A_2^t\,(A_2)_{x_i} & 0 \\ 0 & 0 & -F^tC_i + C_iF \end{pmatrix} \right\} dx_i.$$

Although this does not have the same shape as the Lax pair θ_λ of the $G_{m,n}$-system, we show below that it can be gauged to one. From (3.4), we have

$$dA_1 = A_1 \sum (C_iF^t - FC_i)dx_i + \zeta \sum C_i dx_i \tag{3.6}$$

for some $\zeta : R^n \to \mathcal{M}_{m\times n}$. Thus

$$A_2^t dA_1 = A_2^t \zeta \sum C_i dx_i. \tag{3.7}$$

Since $A^{-1}dA$ is flat and

$$A^{-1}dA = \begin{pmatrix} A_1^t dA_1 & A_1^t dA_2 \\ A_2^t dA_1 & A_2^t dA_2 \end{pmatrix},$$

we have

$$dA_2^t \wedge dA_2 + A_2^t dA_1 \wedge A_1^t dA_2 + A_2^t dA_2 \wedge A_2^t dA_2 = 0. \tag{3.8}$$

By (3.7), $A_1^t dA_2 = (dA_2^t A_1)^t = -(A_2^t dA_1)^t = -\sum C_i\zeta^t A_2 dx_i$. So it follows from (3.8) that $A_2^t dA_2$ is flat, and hence there exists $h : R^n \to O(m-n)$ such that $h^{-1}dh = A_2^t dA_2$. Thus if we do a gauge transform by $\hat{h} = \begin{pmatrix} I & 0 & 0 \\ 0 & h & 0 \\ 0 & 0 & I \end{pmatrix}$ on $g * \theta^I_\lambda$, the resulting connection 1-form is

$$\hat{h} * (g * \theta^I_\lambda) =$$
$$\sum \left\{ \lambda \begin{pmatrix} 0 & 0 & -C_i \\ 0 & 0 & 0 \\ C_i & 0 & 0 \end{pmatrix} + \begin{pmatrix} A_1^t(A_1)_{x_i} & -C_i\zeta^t A_2 h^t & 0 \\ hA_2^t\zeta C_i & 0 & 0 \\ 0 & 0 & -F^tC_i + C_iF \end{pmatrix} \right\} dx_i.$$

Set

$$G = -hA_2^t\zeta.$$

From (3.6), we have $A_1^t(A_1)_{x_i} - (C_iF^t - FC_i) = YC_i$, where $Y = A_1^t\zeta$. Since the left-hand side is skew-symmetric, so is YC_i. But $YC_i = -C_iY$ for all $1 \le i \le n$ implies that $Y = 0$. It follows that $\hat{h} * (g * \theta_\lambda^I)$ is the θ_λ defined by (3.2) with $\xi = \begin{pmatrix} F \\ G \end{pmatrix}$. In other words, ξ is a solution of (3.1). ∎

The $G_{m,n}$-system II is the PDE for

$$(F, G, B) : R^n \to gl_*(n) \times \mathcal{M}_{(m-n)\times n} \times O(n)$$

such that

$$\theta_\lambda^{II} = \sum_{i=1}^n \begin{pmatrix} -FC_i + C_iF^t & C_iG^t & -C_iB^t\lambda \\ -GC_i & 0 & 0 \\ BC_i\lambda & 0 & 0 \end{pmatrix} dx_i \tag{3.9}$$

is flat for all $\lambda \in C$, i.e.,

$$\begin{cases} (f_{ij})_{x_i} + (f_{ji})_{x_j} + \sum_{k=1}^n f_{ki}f_{kj} + \sum_{k=1}^{m-n} g_{ki}g_{kj} = 0, & \text{if } i \neq j, \\ (f_{ij})_{x_k} = f_{ik}f_{kj}, & \text{if } i, j, k \text{ are distinct}, \\ (b_{ij})_{x_k} = f_{kj}b_{ik}, & \text{if } j \neq k, \\ (g_{ij})_{x_k} = f_{kj}g_{ik}, & \text{if } j \neq k. \end{cases} \tag{3.10}$$

As a consequence of Proposition 2.5 we get

3.7 Proposition. *Given $(F, G, B) : R^n \to gl_*(n) \times \mathcal{M}_{(m-n)\times n} \times O(n)$, the following statements are equivalent:*

(i) (F, G, B) is a solution of the $G_{m,n}$-system II (3.10).
(ii) θ_λ^{II} defined by (3.9) is flat for all $\lambda \in C$,
(iii) θ_λ^{II} defined by (3.9) is flat for $\lambda = 1$.

Note that if E is a frame for $v = \begin{pmatrix} F \\ G \end{pmatrix}$, then $E(x, 0) = \begin{pmatrix} A(x) & 0 \\ 0 & B(x) \end{pmatrix}$ and $E^{II} = E\begin{pmatrix} A^{-1} & 0 \\ 0 & I_n \end{pmatrix}$ is a frame for (F, G, B).

If $b_{ji} \neq 0$ for all $1 \le i \le n$, then the third equation of (3.10) implies that $f_{ki} = (b_{ji})_{x_k}/b_{jk}$ if $k \neq i$. In other words, generically system (3.10) depends only on B and G.

3.8 Corollary. *If (F, G, B) is a solution of the $G_{m,n}$-system II (3.10) and $C \in O(n)$ a constant matrix, then (F, G, CB) is also a solution of (3.10). Moreover, if E^{II} is a frame for (F, G, B), then $E^{II}C^{-1}$ is a frame for (F, G, CB).*

Let $U/K = G_{m,n+1}$ with $m \ge n+1$. The rank of U/K is $n+1$. So the corresponding U/K-system has $(n+1)$ independent variables. Below we consider a partial U/K-system of n-variables. Let $(m_1, m_2, m_3, m_4) = (n, m-n, n, 1)$.

We partition a matrix A in $o(m+n+1)$ into 4×4 blocks $A=(A_{ij})$, where $A_{ij}\in\mathcal{M}_{m_i\times m_j}$. Let

$$a_i=\begin{pmatrix}0&0&-C_i&0\\0&0&0&0\\C_i&0&0&0\\0&0&0&0\end{pmatrix},\ \text{where } C_i=\text{diag}(0,\ldots,1,\ldots,0) \text{ as before.}$$

Then the space $\mathcal{A}$ spanned by $a_1,\cdots,a_n$ is an n-dimensional abelian subspace of $\mathcal{P}$. Let $\mathcal{K}_{\mathcal{A}}$ and $\mathcal{U}_{\mathcal{A}}$ denote the centralizer of $\mathcal{A}$ in $\mathcal{K}$ and $\mathcal{U}$ respectively. Then $\mathcal{P}\cap\mathcal{U}_{\mathcal{A}}^{\perp}$ is the space of elements of the form

$$v=\begin{pmatrix}0&0&F&b\\0&0&G&0\\-F^t&-G^t&0&0\\-b^t&0&0&0\end{pmatrix},$$

where, $F\in gl(n)_*$, $b\in\mathcal{M}_{n\times 1}$, $G\in\mathcal{M}_{(m-n)\times n}$.

The *partial* $G_{m,n+1}$*-system* is the PDE for the map $(F,G,b):R^n\to gl(n)_*\times\mathcal{M}_{(m-n)\times n}\times\mathcal{M}_{n\times 1}$ such that

$$\begin{aligned}\Theta_\lambda=&\sum_i\lambda\begin{pmatrix}0&0&-C_i&0\\0&0&0&0\\C_i&0&0&0\\0&0&0&0\end{pmatrix}dx_i\\&+\sum_i\begin{pmatrix}-FC_i+C_iF^t&C_iG^t&0&0\\-GC_i&0&0&0\\0&0&-F^tC_i+C_iF&C_ib\\0&0&-b^tC_i&0\end{pmatrix}dx_i\end{aligned}\tag{3.11}$$

is flat for all $\lambda\in C$, i.e.,

$$\begin{cases}(f_{ij})_{x_i}+(f_{ji})_{x_j}+\sum_k f_{ki}f_{kj}+\sum_k g_{ki}g_{kj}=0, & \text{if } i\neq j,\\(f_{ij})_{x_k}=f_{ik}f_{kj}, & \text{if } i,j,k \text{ distinct},\\(g_{ij})_{x_k}=g_{ik}f_{kj}, & \text{if } k\neq j,\\(f_{ij})_{x_j}+(f_{ji})_{x_i}+\sum_k f_{ik}f_{jk}+b_ib_j=0, & \text{if } i\neq j,\\(b_i)_{x_j}=f_{ij}b_j.\end{cases}\tag{3.12}$$

The *partial* $G_{m,n+1}$*-system I* is the PDE for maps

$$(A_1,F,b):R^n\to\mathcal{M}^0_{m\times n}\times gl_*(n)\times\mathcal{M}_{n\times 1}$$

such that

$$\Theta^I_\lambda=\sum_i\begin{pmatrix}0&-A_1C_i\lambda&0\\C_iA_1^t\lambda&-F^tC_i+C_iF&C_ib\\0&-b^tC_i&0\end{pmatrix}dx_i\tag{3.13}$$

is flat for all $\lambda\in C$, i.e.,

$$\begin{cases}(f_{ij})_{x_j}+(f_{ji})_{x_i}+\sum_{k=1}^n f_{ik}f_{jk}+b_ib_j=0, & i\neq j,\\(f_{ij})_{x_k}=f_{ik}f_{kj}, & i,j,k \text{ distinct},\\(b_i)_{x_j}=f_{ij}b_j, & i\neq j,\\(a_{ki})_{x_j}=f_{ij}a_{kj}, & i\neq j.\end{cases}\tag{3.14}$$

By Proposition 2.5 we have

3.9 Proposition. *Given a map $(A_1, F, b) : R^n \to \mathcal{M}^0_{m\times n} \times gl_*(n) \times \mathcal{M}_{n\times 1}$, the following statements are equivalent:*
(i) (A_1, F, b) is a solution of the partial $G_{m,n+1}$-system I (3.14),
(ii) Θ^I_λ defined by (3.13) is flat for all $\lambda \in C$,
(iii) Θ^I_λ defined by (3.13) is flat for $\lambda = 1$.

4. $G^1_{m,n}$-Systems

In this section, we assume $m \geq n+1$ and

$$U/K = G^1_{m,n} = O(m+n,1)/(O(m)\times O(n,1)).$$

We write down the $G^1_{m,n}$-system I and II explicitly.

Let $\mathcal{U} = o(m+n,1) = \{X \in gl(m+n+1) \mid X^t I_{m+n,1} + I_{m+n,1}X = 0\}$ and $\sigma : \mathcal{U} \to \mathcal{U}$ be an involution defined by $\sigma(X) = I^{-1}_{m,n+1} X I_{m,n+1}$, where $I_{p,q} = \begin{pmatrix} I_p & 0 \\ 0 & -I_q \end{pmatrix}$. Let $\mathcal{K}$ and $\mathcal{P}$ denote the $+1$ and -1 eigenspaces of σ respectively. Then the Cartan decomposition is $\mathcal{U} = \mathcal{K} + \mathcal{P}$, where

$$\mathcal{K} = o(m)\times o(n,1) = \left\{ \begin{pmatrix} Y_1 & 0 \\ 0 & Y_2 \end{pmatrix} \middle| \, Y_1 \in o(m), Y_2 \in o(n,1) \right\},$$
$$\mathcal{P} = \left\{ \begin{pmatrix} 0 & \xi \\ -J\xi^t & 0 \end{pmatrix} \middle| \, \xi \in \mathcal{M}_{m\times(n+1)} \right\}.$$

Here $\mathcal{M}_{m\times n}$ is the set of $m\times n$ matrices and $J = I_{n,1} = \mathrm{diag}(1,\cdots,1,-1)$. It is easy to see that

$$\mathcal{A} = \left\{ \begin{pmatrix} 0 & -DJ \\ D^t & 0 \end{pmatrix} \middle| \, D = (d_{ij}) \in \mathcal{M}_{m\times(n+1)},\ d_{ij} = 0 \text{ if } i \neq j \right\}$$

is a maximal abelian subalgebra in $\mathcal{P}$. Let

$$a_i = \begin{pmatrix} 0 & -D_iJ \\ D_i^t & 0 \end{pmatrix},$$

where $D_i \in \mathcal{M}_{m\times(n+1)}$ is the matrix all whose entries are zero except that the (i,i)-th entry is equal to 1. Then $a_1, \ldots, a_{n+1}$ form a basis of $\mathcal{A}$. The U/K-system (1.1) for this symmetric space is the PDE for $\xi = (\xi_{ij}) : R^{n+1} \to \mathcal{M}_{m\times(n+1)}$ with $\xi_{ii} = 0$ for all $1 \leq i \leq n+1$ such that

$$\theta_\lambda = \sum_i \left\{ \lambda \begin{pmatrix} 0 & -D_iJ \\ D_i^t & 0 \end{pmatrix} + \begin{pmatrix} D_i\xi^t - \xi D_i^t & 0 \\ 0 & -J\xi^t D_i J + D_i^t\xi \end{pmatrix} \right\} dx_i \qquad (4.1)$$

is a family of flat connections on R^{n+1} for all $\lambda \in C$, i.e.,

$$\begin{cases} D_i\xi^t_{x_j} - \xi_{x_j}D^t_i - D_j\xi^t_{x_i} + \xi_{x_i}D^t_j = [D_i\xi^t - \xi D^t_i, D_j\xi^t - \xi D^t_j], & i \neq j, \\ D^t_i\xi_{x_j} - J\xi^t_{x_j}D_iJ - D^t_j\xi_{x_i} + J\xi^t_{x_i}D_jJ & \\ \qquad = [D^t_i\xi - J\xi^tD_iJ, D^t_j\xi - J\xi^tD_jJ]. & i \neq j. \end{cases} \tag{4.2}$$

Let

$$gl(n+1)_* = \{(x_{ij}) \in gl(n+1) \mid x_{ii} = 0,\ 1 \leq i \leq n+1\}.$$

The $G^1_{m,n}$-system I is the PDE for $(A_1, F) : R^{n+1} \to \mathcal{M}^0_{m\times(n+1)} \times gl_*(n+1)$ such that

$$\theta^I_\lambda = \sum_{i=1}^n \begin{pmatrix} 0 & -A_1C_iJ\lambda \\ C_iA^t_1\lambda & -JF^tC_iJ + C_iF \end{pmatrix} dx_i \tag{4.3}$$

is a flat connection on R^{n+1} for all $\lambda \in C$, i.e.,

$$\begin{cases} -(A_1)_{x_j}C_iJ + (A_1)_{x_i}C_jJ = -A_1C_iJ\eta_j + A_1C_jJ\eta_i, \\ (\eta_i)_{x_j} - (\eta_j)_{x_i} = [\eta_i, \eta_j], \end{cases}$$
$$\text{where} \quad \eta_i = -JF^tC_iJ + C_iF. \tag{4.4}$$

Next we write down the U/K-system II. Write

$$\xi = \begin{pmatrix} F \\ G \end{pmatrix}, \quad D_i = \begin{pmatrix} C_i \\ 0 \end{pmatrix}.$$

Then the $G^1_{m,n}$-system II is the PDE for

$$(F, G, B) : R^{n+1} \to gl_*(n+1) \times \mathcal{M}_{(m-n-1)\times(n+1)} \times O(n,1)$$

such that

$$\theta^{II}_\lambda = \sum_{i=1}^{n+1} \begin{pmatrix} -FC_i + C_iF^t & C_iG^t & -C_iB^tJ\lambda \\ -GC_i & 0 & 0 \\ BC_i\lambda & 0 & 0 \end{pmatrix} dx_i \tag{4.5}$$

is flat for all $\lambda \in C$, i.e.,

$$\begin{cases} (f_{ij})_{x_i} + (f_{ji})_{x_j} + \sum_{k=1}^{n+1} f_{ki}f_{kj} + \sum_{k=1}^{m-n-1} g_{ki}g_{kj} = 0, & \text{if } i \neq j, \\ (f_{ij})_{x_k} = f_{ik}f_{kj}, & \text{if } i, j, k \text{ are distinct}, \\ (b_{ij})_{x_k} = f_{kj}b_{ik}, & \text{if } j \neq k, \\ (g_{ij})_{x_k} = f_{kj}g_{ik}, & \text{if } j \neq k. \end{cases} \tag{4.6}$$

It follows from Proposition 2.5 that

4.1 Proposition. *The following statements are equivalent for maps (F, G, B) : $R^{n+1} \to gl_*(n+1) \times \mathcal{M}_{(m-n-1)\times(n+1)} \times O(n,1)$:*

(i) It is a solution of the $G^1_{m,n}$-system II (4.6).

(ii) θ^{II}_λ defined by (4.5) is a flat connection on R^{n+1} for all $\lambda \in C$,

(iii) θ^{II}_λ defined by (4.5) is flat for $\lambda = 1$.

Next we write down the partial $G^1_{m,n}$-systems of n variables. Let $(m_1, m_2, m_3, m_4) = (n, m-n, n, 1)$. We partition a matrix A in $o(m+n,1)$ into 4×4 blocks $A = (A_{ij})$, where $A_{ij} \in \mathcal{M}_{m_i\times m_j}$. Let

$$a_i = \begin{pmatrix} 0 & 0 & -C_i & 0 \\ 0 & 0 & 0 & 0 \\ C_i & 0 & 0 & 0 \\ 0 & 0 & 0 & 0 \end{pmatrix}, \text{ where } C_i = \operatorname{diag}(0, \ldots, 1, \ldots, 0) \text{ as before.}$$

Then the space $\mathcal{A}$ spanned by $a_1, \cdots, a_n$ is an n-dimensional abelian subspace of $\mathcal{P}$ (it is not a maximal one). Let $\mathcal{K}_{\mathcal{A}}$ and $\mathcal{U}_{\mathcal{A}}$ denote the centralizer of $\mathcal{A}$ in $\mathcal{K}$ and $\mathcal{U}$ respectively. Then $\mathcal{P}\cap\mathcal{U}_{\mathcal{A}}^\perp$ is the space of elements of the form

$$v = \begin{pmatrix} 0 & 0 & F & b \\ 0 & 0 & G & 0 \\ -F^t & -G^t & 0 & 0 \\ b^t & 0 & 0 & 0 \end{pmatrix},$$

where, $F \in gl(n)_*$, $b \in \mathcal{M}_{n\times 1}$, and $G \in \mathcal{M}_{(m-n)\times n}$.

The partial $G^1_{m,n}$-system is the system for (F, G, b) such that

$$\begin{aligned} \tau_\lambda = &\sum_i \lambda \begin{pmatrix} 0 & 0 & -C_i & 0 \\ 0 & 0 & 0 & 0 \\ C_i & 0 & 0 & 0 \\ 0 & 0 & 0 & 0 \end{pmatrix} dx_i \\ &+ \sum_i \begin{pmatrix} -FC_i + C_iF^t & C_iG^t & 0 & 0 \\ -GC_i & 0 & 0 & 0 \\ 0 & 0 & -F^tC_i + C_iF & C_ib \\ 0 & 0 & b^tC_i & 0 \end{pmatrix} dx_i, \end{aligned} \tag{4.7}$$

is flat for all $\lambda \in C$, i.e.,

$$\begin{cases} (f_{ij})_{x_i} + (f_{ji})_{x_j} + \sum_k f_{ki}f_{kj} + \sum_k g_{ki}g_{kj} = 0, & \text{if } i \neq j, \\ (f_{ij})_{x_k} = f_{ik}f_{kj}, & \text{if } i, j, k \text{ distinct}, \\ (g_{ij})_{x_k} = g_{ik}f_{kj}, & \text{if } j \neq k, \\ (f_{ij})_{x_j} + (f_{ji})_{x_i} + \sum_k f_{ik}f_{jk} - b_ib_j = 0, & \text{if } i \neq j, \\ (b_i)_{x_j} = f_{ij}b_j, & \text{if } i \neq j. \end{cases} \tag{4.8}$$

The partial $G^1_{m,n}$-system I is the PDE for the map $(A_1, F, b) : R^n \to gl(n)_* \times \mathcal{M}_{(m-n)\times n} \times \mathcal{M}_{n\times 1}$ such that

$$\tau^I_\lambda = \sum_i \begin{pmatrix} 0 & -A_1C_i\lambda & 0 \\ C_iA_1^t\lambda & -F^tC_i + C_iF & C_ib \\ 0 & b^tC_i & 0 \end{pmatrix} dx_i \tag{4.9}$$

is flat for all $\lambda \in C$. It follows from Proposition 2.5 that

4.2 Proposition. *The following statements are equivalent for maps* $(A_1, F, b) : R^n \to \mathcal{M}^0_{m\times n} \times gl_*(n) \times \mathcal{M}_{n\times 1}$*:*
(i) It is a solution of the partial $G^1_{m,n}$*-system I:*

$$\begin{cases} (f_{ij})_{x_j} + (f_{ji})_{x_i} + \sum_{k=1}^n f_{ik}f_{jk} - b_ib_j = 0, & i \neq j, \\ (f_{ij})_{x_k} = f_{ik}f_{kj}, & i,j,k \text{ distinct}, \\ (b_i)_{x_j} = f_{ij}b_j, & i \neq j, \\ (a_{ki})_{x_j} = f_{ij}a_{kj}, & i \neq j. \end{cases} \tag{4.10}$$

(ii) τ^I_λ *defined by (4.9) is a flat connection on* R^n *for all* $\lambda \in C$.
(ii) τ^I_λ *defined by (4.9) is a flat connection on* R^n *for* $\lambda = 1$.

5. Moving frame method for submanifolds

In this section, we review some elementary theory of submanifolds in Euclidean space (cf. Chapter 2 of [PT]). The basic local invariants of submanifolds in R^N are the first, second fundamental forms and the induced normal connection. These invariants satisfy the Gauss, Codazzi and Ricci equations. If we use the method of moving frames, then these equations arise as the condition for a $o(N)$-valued 1-form to be flat. Natural geometric conditions on submanifolds in R^N often lead to interesting PDE. If in addition these submanifolds come in a family that depends holomorphically on a parameter, then the corresponding PDE has a Lax connection. Therefore we can use techniques from soliton theory to study these submanifolds.

Let $X : M^n \to R^{n+m}$ be an immersed submanifold. Henceforth we agree on the following index conventions unless otherwise stated:

$$1 \leq i,j,k \leq n, \quad n+1 \leq \alpha, \beta, \gamma \leq n+m, \quad 1 \leq A, B, C \leq n+m.$$

Let e_A be a local orthonormal frame on M such that e_α are normal field, and

$$E = (e_1, \cdots, e_{n+m}),$$

i.e., the A-th column of E is e_A. Thus $E \in O(m+n)$. Let ω_i be the local orthonormal frame of T^*M dual to e_i. Then

$$dX = \sum_i \omega_i e_i,$$

and the first fundamental form is

$$I = \sum_i \omega_i^2.$$

Set $\omega_{AB} = \langle e_A, de_B \rangle$, where $\langle , \rangle$ is the inner product on R^{n+m}, or equivalently,

$$de_B = \sum_A \omega_{AB} e_A = -\sum_A \omega_{BA} e_A.$$

Write this in matrix form to get

$$dE = E(\omega_{AB}),$$

i.e.,

$$(\omega_{AB}) = E^{-1}dE.$$

In other words, (ω_{AB}) is a flat $o(n+m)$-valued 1-form on M, and ω_{AB} satisfies the Maurer-Cartan equation:

$$d\omega_{AB} = -\sum_C \omega_{AC} \wedge \omega_{CB}. \tag{5.1}$$

The structure equation is

$$d\omega_i = -\sum_j \omega_{ij} \wedge \omega_j, \quad \omega_{ij} + \omega_{ji} = 0. \tag{5.2}$$

Here (ω_{ij}) is the Levi-Civita connection 1-form for the induced metric I, and it can be computed in terms of $\omega_1, \cdots, \omega_n$ by solving the structure equation (5.2).

The second fundamental form of M is

$$II = \sum_{i,\alpha} \omega_{i\alpha}\omega_i e_\alpha.$$

Given $\xi \in \nu(M)_x$, the *shape operator* A_ξ is the self-adjoint operator defined by $< II, \xi >$, i.e.,

$$< II(v_1, v_2), \xi >=< A_\xi(v_1), v_2 > .$$

The eigenvalues and eigendirections of A_ξ are the *principal curvatures* and *principal directions* of M with respect to ξ respectively.

The induced connection $\nabla^\perp$ on the normal bundle $\nu(M)$ is defined by

$$\nabla^\perp \xi = (d\xi)^\perp$$

for any normal field ξ, where $(d\xi)^\perp$ is the normal component of $d\xi$. In particular,

$$\nabla^\perp e_\alpha = -\sum_\beta \omega_{\alpha\beta} e_\beta.$$

The normal curvature is

$$\Omega^\perp_{\alpha\beta} = d\omega_{\alpha\beta} + \sum_\gamma \omega_{\alpha\gamma} \wedge \omega_{\gamma\beta}.$$

Equation (5.1) gives the fundamental equations for M: The (i,j)-th entry gives the Gauss equation

$$\begin{aligned} d\omega_{ij} &= -\sum_k \omega_{ik} \wedge \omega_{kj} - \sum_\alpha \omega_{i\alpha} \wedge \omega_{\alpha j}, \\ &= -\sum_k \omega_{ik} \wedge \omega_{kj} + \Omega_{ij}, \end{aligned} \tag{5.3}$$

where $\Omega_{ij} = \sum_\alpha \omega_{i\alpha} \wedge \omega_{j\alpha}$ is the Riemann curvature tensor of I. The $i\alpha$-th entry gives the Codazzi equations

$$d\omega_{i\alpha} = -\sum_j \omega_{ij} \wedge \omega_{j\alpha} - \sum_\beta \omega_{i\beta} \wedge \omega_{\beta\alpha}, \tag{5.4}$$

and the $\alpha\beta$-th entry gives the Ricci's equations:

$$d\omega_{\alpha\beta} + \sum_\gamma \omega_{\alpha\gamma} \wedge \omega_{\gamma\beta} = \Omega^\perp_{\alpha\beta} = -\sum_i \omega_{\alpha i} \wedge \omega_{i\beta}.$$

5.1 Fundamental Theorem of Submanifolds in Euclidean space. *Let (M, g) be an n-dimensional Riemannian manifold, ξ a rank m orthogonal vector bundle over M, ∇_0 an $O(m)$-connection on ξ, and b a smooth section of $S^2(T^*M) \otimes \xi$. If g, b and ∇_0 satisfy the Gauss-Codazzi-Ricci equations, then there exist a local isometric immersion of M into R^{n+m} and a bundle isomorphism between ξ and $\nu(M)$ such that g, b and ∇_0 are the first, second fundamental forms and induced normal connection respectively. This immersion is unique up to rigid motions.*

A normal vector field η is *parallel* if $\nabla^\perp \eta = 0$. The normal bundle $\nu(M)$ is *flat* if the induced normal connection $\nabla^\perp$ is flat, i.e.,

$$\Omega^\perp_{\alpha\beta} = d\omega_{\alpha\beta} + \sum_\gamma \omega_{\alpha\gamma} \wedge \omega_{\gamma\beta} = 0. \tag{5.5}$$

If $\nu(M)$ is flat, then it follows from (5.5) that there exists a local parallel normal frame. So we may assume that $(e_{n+1}, \cdots, e_{n+m})$ is parallel, i.e.,

$$\omega_{\alpha\beta} = 0.$$

Then the Ricci equation is

$$0 = d\omega_{\alpha\beta} = \sum_i \omega_{i\alpha} \wedge \omega_{i\beta}. \tag{5.6}$$

In particular, this implies that $[A_{e_\alpha}, A_{e_\beta}] = 0$ for all α, β. Hence the family $\{A_v \mid v \in \nu(M)_p\}$ of shape operators of M at p is a family of commuting self-adjoint operators on TM_p, and generically there is a smooth common eigenframe.

Henceforth, we assume that $\nu(M)$ is flat, and e_α is parallel, i.e., $\omega_{\alpha\beta} = 0$. So equation (5.4) becomes

$$d\omega_{i\alpha} = -\sum_j \omega_{ij} \wedge \omega_{j\alpha}, \quad \omega_{ij} + \omega_{ji} = 0. \tag{5.7}$$

Next we recall the well-known Cartan Lemma.

5.2 Cartan Lemma. *If $\tau_1, \cdots, \tau_n$ are linearly independent 1-forms on an n-dimensional manifold, then there exists a unique $o(n)$-valued 1-form (τ_{ij}) such that*

$$d\tau_i = -\sum_{j=1}^{n} \tau_{ij} \wedge \tau_j, \quad \tau_{ij} + \tau_{ji} = 0, \quad 1 \leq i \leq n. \tag{5.8}$$

A direct computation gives

5.3 Corollary. *If $\tau_i = b_i dx_i$, then the solution (τ_{ij}) for system (5.8) is given by*

$$\tau_{ij} = \frac{(b_i)_{x_j}}{b_j} dx_i - \frac{(b_j)_{x_i}}{b_i} dx_j.$$

5.4 Definition. Let M^n be a submanifold in a Riemannian manifold N^{n+m}. The normal bundle $\nu(M)$ is called *non-degenerate* if the space of shape operators $\{A_v \mid v \in \nu(M)_p\}$ has dimension n for all $p \in M$.

When M is a surface in R^3, it is well-known that if $p \in M$ is not umbilic then there exist a local coordinates x_1, x_2 such that $\frac{\partial}{\partial x_1}, \frac{\partial}{\partial x_2}$ are eigenvectors for the shape operator A_{e_3}. In other words, the two fundamental forms are of the form

$$I = a_1^2 dx_1^2 + a_2^2 dx_2^2, \quad II = b_1 dx_1^2 + b_2 dx_2^2$$

for some functions a_i, b_i. Such coordinates are called *line of curvature coordinates*. We generalize this notion to submanifolds below.

5.5 Definition. Let M be an n-dimensional submanifold of R^m with flat normal bundle, and $\{e_\alpha\} = \{e_{n+1}, \cdots, e_m\}$ a local orthonormal parallel normal frame. Local coordinates $x_1, \cdots, x_n$ are called *line of curvature coordinates* with respect to $\{e_\alpha\}$ if $\{\frac{\partial}{\partial x_i}\}$ is a common orthogonal eigenbasis for the shape operators A_v for all $v \in \nu(M)$, or equivalently, if the two fundamental forms of M are of the form

$$I = \sum_{i=1}^{n} a_i^2 dx_i^2,$$

$$II = \sum_{i=1, j=1}^{n, m-n} b_{ji} dx_i^2 e_{n+j}$$

for some smooth functions a_i and b_{ij}. If in addition $a_1^2 + \cdots + a_n^2 = 1$ ($a_1^2 + \cdots + a_{n-1}^2 - a_n^2 = \pm 1$ resp.), then $x_1, \cdots, x_n$ are called *spherical (hyperbolic resp.) line of curvature coordinates.*

Let M be a submanifold of R^{n+m} with flat normal bundle. Although generically there exist orthonormal tangent frame e_i that are common eigenbasis for all shape operators, we can not always find coordinates $x_1, \cdots, x_n$ so that $\frac{\partial}{\partial x_i}$ is parallel to e_i for all i. However, n-dimensional submanifolds in R^{2n-1} with constant sectional curvature are known to have flat and non-degenerate normal bundle, and admit spherical line of curvature coordinates (cf. [Ca], [M]).

6. Submanifolds associated to $G_{m,n}$-systems

In this section, we describe submanifolds associated to the $G_{m,n}$-systems I and II.

Recall that the Lax connection of the $G_{m,n}$- system I (3.4) is θ^I_λ defined by (3.3). Let θ_{ij} denote the (i,j)-th entry of θ^I_1 (i.e., θ^I_λ at $\lambda = 1$). Then

$$\theta_{ij} = \begin{cases} 0, & \text{if } 1 \le i,j \le m, \\ a_{j,i-m} dx_{i-m}, & \text{if } m+1 \le i \le m+n, 1 \le j \le m, \\ -f_{j-m,i-m} dx_{j-m} + f_{i-m,j-m} dx_{i-m}, & \text{if } m+1 \le i,j \le m+n. \end{cases}$$

Set

$$\begin{aligned} &\omega_i = \theta_{m+i,1} = a_{1i} dx_i, \quad \omega_{ij} = \theta_{m+i,m+j} = -f_{ji} dx_j + f_{ij} dx_i, \text{ if } 1 \le i,j \le n, \\ &\omega_{n+i,n+j} = \theta_{ij} = 0, \quad \text{if} \quad 1 \le i,j \le m, \\ &\omega_{i,n+j} = \theta_{m+i,j} = a_{ji} dx_i, \quad \text{if} \quad 1 \le i \le n, 1 \le j \le m. \end{aligned}$$

Since $\theta^I_1 = (\theta_{ij})$ is flat, we have

$$\begin{aligned} d\omega_i &= d\theta_{m+i,1} \\ &= -\sum_{j=1}^m \theta_{m+i,j} \wedge \theta_{j1} - \sum_{j=1}^n \theta_{m+i,m+j} \wedge \theta_{m+j,1} \\ &= -\sum_{j=1}^n \theta_{m+i,m+j} \wedge \theta_{m+j,1} = -\sum_{j=1}^n \omega_{ij} \wedge \omega_j. \end{aligned}$$

If $\sum_{i=1}^n \omega_i^2$ is non-degenerate, then (ω_{ij}) is the Levi-Civita connection 1-form for the metric $\sum_{i=1}^n \omega_i^2$. Let E be a trivialization of θ^I_1, i.e., $E^{-1}dE = \theta^I_1$. Let e_{n+i} denote the i-th column of E for $1 \le i \le m$, and let e_i denote the $(m+i)$-th column of E, i.e., $E = (e_{n+1}, \cdots, e_{n+m}, e_1, \cdots, e_n)$. It follows from $dE = E\theta^I_1$ that e_{n+1} is an n-dimensional submanifold in $S^{n+m-1} \subset R^{n+m}$ such that $e_{n+1}, e_{n+2}, \cdots, e_{n+m}$ is a parallel normal frame and $(x_1, \cdots, x_n)$ is a spherical line of curvature coordinates such that the two fundamental forms are

$$I = \sum_{i=1}^n a_{1i}^2 dx_i^2, \quad II = \sum_{i=1,\alpha=n+1}^{n,m+n} a_{1i} a_{\alpha-n,i} dx_i^2 e_\alpha.$$

We state this and the converse in the Theorem below.

6.1 Theorem. *(Flat n-submanifolds in S^{m+n-1}). Let X be a local isometric immersion of flat n-dimensional submanifold in S^{n+m-1} with flat, non-degenerate normal bundle, and $m \ge n$. Let $e_{n+1} = X$, and fix a local parallel normal frame $e_{n+2}, \cdots, e_{n+m}$. Then:*

(i) There exist line of curvature coordinates $(x_1, ..., x_n)$ and a smooth $\mathcal{M}_{m\times n}$-valued map $A_1 = (a_{ij})$ such that $A_1^t A_1 = I$ and the first and second fundamental forms of X are given by

$$I = \sum_{i=1}^n a_{1i}^2 \, dx_i^2, \qquad II = \sum_{i=1}^n \sum_{j=2}^m a_{1i} a_{ji} dx_i^2 \, e_{n+j}. \tag{6.1}$$

(ii) Let $f_{ij} = (a_{1i})_{x_j}/a_{1j}$ for $i \neq j$, $f_{ii} = 0$ for all $1 \leq i, j \leq n$, and $F = (f_{ij}) \in gl_*(n)$. Then the Gauss, Codazzi and Ricci equations for the immersion X is the $G_{m,n}$-system I (3.4) for (A_1, F). In other words, (A_1, F) is a solution of (3.4).

(iii) Let $e_i = \frac{1}{a_{1i}} X_{x_i}$, $g = (e_{n+1}, \cdots, e_{m+n}, e_1, \cdots, e_n) \in O(m+n)$. Then

$$g^{-1}dg = \sum_i \begin{pmatrix} 0 & -A_1 C_i \\ C_i A_1^t & -F^t C_i + C_i F \end{pmatrix} dx_i, \tag{6.2}$$

which is equal to the Lax connection (3.3) θ_λ^I of equation (3.4) at $\lambda = 1$.

(iv) Conversely, if (A_1, F) is a solution of (3.4), then system (6.2) is solvable. Let g be a solution of (6.2), and X the first column g. If all entries of the first row of A_1 are non-zero, then X is an isometric immersion of flat n-submanifolds in S^{n+m-1} with flat and non-degenerate normal bundle such that the two fundamental forms are as in (i), where $A_1 = (a_{ij})$.

PROOF.

(i) can be proved the same way as for isometric immersions of n-submanifolds in R^{2n-1} with constant sectional curvature -1 (cf. [Ca], [M]).

From (i), we have

$$\omega_i = a_{1i} dx_i, \quad \omega_{i,n+j} = a_{ji} dx_i, \quad \omega_{\alpha\beta} = 0.$$

By Corollary 5.3, we get $\omega_{ij} = f_{ij} dx_i - f_{ji} dx_j$. So $g^{-1}dg = (\omega_{AB})$, and (ii), (iii) follow.

(iv) follows from the Fundamental Theorem of submanifolds in Euclidean space.

∎

When $m = n$, the above theorem was proved by Tenenblat in [Ten], where equation (3.4) is called the *generalized wave equation.*

If we use a different parallel frame $\tilde{e}_{n+2}, \cdots, \tilde{e}_{n+m}$ for the immersion X in the above Theorem, then there exists a constant matrix $p = (p_{ij}) \in O(m-1)$ such that $\tilde{e}_{n+i} = \sum_{j=2}^m p_{ij} e_{n+j}$ for $2 \leq i \leq m$. The solution of (3.4) given by X and parallel frame $\tilde{e}_\alpha$ is $(\tilde{p}A_1, F)$, where $\tilde{p} = \begin{pmatrix} 1 & 0 \\ 0 & p \end{pmatrix}$ and $p = (p_{ij})$.

Suppose (A, F) is a solution of (3.4), and $p \in O(m)$. By Corollary 3.5, (pA, F) is also a solution of (3.4). As a consequence of Theorem 6.1 (iv), we get

6.2 Corollary. *Let $X, e_{n+2}, \cdots, e_{n+m}$, and (A_1, F) be as in Theorem 6.1, and $c = (c_1, \cdots, c_m)$ a unit vector. If all components of cA_1 never vanish, then $Y = c_1 X + c_2 e_{n+2} + \cdots + c_m e_{n+m}$ is again an immersion of a flat n-submanifold in S^{n+m-1} with flat and non-degenerate normal bundle.*

6.3 Theorem. *(Flat n-submanifolds in R^{n+m}). Let $n \leq m$, and X a local isometric immersion of flat n-dimensional submanifold in R^{m+n} with flat and non-degenerate normal bundle. Fix a local parallel normal frame $e_{n+1}, \cdots, e_{n+m}$. Then:*

(i) *There exist a line of curvature coordinates* $(x_1, ..., x_n)$, *a smooth* $\mathcal{M}_{m\times n}$*-valued map* $A_1 = (a_{ij})$ *and a smooth* $\mathcal{M}_{1\times n}$*-valued map* $b = (b_1, ..., b_n)$ *such that* $A_1^t A_1 = I$ *and the first and second fundamental forms of* X *are given by*

$$I = \sum_i b_i^2 dx_i^2, \quad II = \sum_{i=1}^n \sum_{j=1}^m b_i a_{ji} dx_i^2 e_{n+j}. \tag{6.3}$$

(ii) *Let* $f_{ij} = (b_i)_{x_j}/b_j$ *for* $i \neq j$, $f_{ii} = 0$, *and* $F = (f_{ij})$. *Then* (A_1, F) *is a solution of the* $G_{m,n}$*-system I (3.4). Moreover, if*

$$e_i := \frac{1}{b_i} X_{x_i}, \quad g := (e_{n+1}, \cdots, e_{n+m}, e_1, \cdots, e_n),$$

then

$$g^{-1}dg = \sum_i \begin{pmatrix} 0 & -A_1 C_i \\ C_i A_1^t & -F^t C_i + C_i F \end{pmatrix} dx_i,$$

which is the Lax connection θ_λ^I *defined by (3.3) for equation (3.4) at* $\lambda = 1$.

(iii) *Conversely, if* (A_1, F) *is a solution of (3.4) and* $b_1, \cdots, b_n$ *satisfy*

$$(b_i)_{x_j} = f_{ij} b_j, \quad 1 \leq i \neq j \leq n, \tag{6.4}$$

then there exists a local isometric immersion of R^n *in* R^{n+m} *such that the two fundamental forms are given by (6.3), where* $A_1 = (a_{ij})$ *and* $F = (f_{ij})$.

(iv) X *lies in a hypersphere of radius 1 if and only if* $b = vA_1$ *for some constant unit vector* $v \in \mathcal{M}_{1\times m}$.

PROOF. Statements (i), (ii) and (iii) follow from an argument similar to those for Theorem 6.1. To prove (iv), we assume $\|X - X_0\| = 1$ for some constant vector X_0. Then $X - X_0$ is a parallel normal field, which implies that there exists a constant unit vector $v = (v_1, \cdots, v_m)$ such that $X - X_0 = \sum_i v_i e_{n+i}$. Hence

$$d(X - X_0) = dX = \sum_i v_i de_{n+i} = \sum_i v_i \omega_{j,n+i} e_j = \sum_i v_i a_{ij} dx_j e_j.$$

But $dX = \sum_j b_j dx_j e_j$. So $b = vA_1$. The converse can be proved by reversing the argument. ∎

6.4 Remark. System (6.4) was studied by Darboux [Da]. It can be solved if and only if

$$(f_{ij})_{x_k} = f_{ik} f_{kj}, \quad i, j, k \text{ distinct}. \tag{6.5}$$

This condition is obtained by equating $(b_i)_{x_j x_k} = (b_i)_{x_k x_j}$:

$$\begin{aligned}(b_i)_{x_j x_k} &= (f_{ij} b_j)_{x_k} = (f_{ij})_{x_k} b_j + f_{ij} f_{jk} b_k \\ &= (f_{ik} b_k)_{x_j} = (f_{ik})_{x_j} b_k + f_{ik} f_{kj} b_j.\end{aligned}$$

Moreover, the space of solutions of (6.4) depends on n arbitrary smooth functions of one variable. In fact, if $\xi(t) = (\xi_1(t), \cdots, \xi_n(t))$ is a curve from $(-\epsilon, \epsilon)$ to R^n such that $\xi_i'(t)$ never vanishes for all $1 \leq i \leq n$, then given any smooth maps $b_1^0, \cdots, b_n^0 : (-\epsilon, \epsilon) \to R$ there exists a unique solution $(b_1, \cdots, b_n)$ of (6.4) such that $b_i(\xi(t)) = b_i^0(t)$. The compatibility condition (6.5) is the third equation of (3.4). So $(b_1, \cdots, b_n)$ in Theorem 6.3 (iii) always exists.

We can use a proof similar to that of the previous theorem to deduce:

6.5 Theorem. *(Local isometric immersion of S^n in S^{n+m}). Suppose X is a local isometric immersion of S^n in S^{n+m} with flat and non-degenerate normal bundle, and $\{e_\alpha\}$ is a parallel normal frame. Then:*

(i) There exist a local coordinate system $(x_1, \ldots, x_n)$, a smooth $\mathcal{M}^0_{m\times n}$-valued map $A_1 = (a_{ij})$ and a smooth $\mathcal{M}_{n\times 1}$-valued map $b = (b_1, \cdots, b_n)^t$ such that the two fundamental forms of X are given by

$$I = \sum_i b_i^2\, dx_i^2, \quad II = \sum_{i=1,j=1}^{n,m} a_{ji} b_i dx_i^2 e_{n+j}. \tag{6.6}$$

(ii) Let $f_{ij} = (b_i)_{x_j}/b_j$ if $i \neq j$, $f_{ii} = 0$ for all $1 \leq i \leq n$, and $F = (f_{ij})$. Then (A_1, F, b) is a solution of the partial $G_{m,n+1}$-system I (3.14).

(iii) Let $e_i = \frac{1}{b_i} X_{x_i}$, and $g = (e_{n+1}, \cdots, e_{n+m}, e_1, \cdots, e_n, X)$. Then

$$g^{-1}dg = \sum_i \begin{pmatrix} 0 & -A_1 C_i & 0 \\ C_i A_1^t & -F^t C_i + C_i F & C_i b \\ 0 & -b^t C_i & 0 \end{pmatrix} dx_i, \tag{6.7}$$

which is the Lax connection Θ_λ defined in (3.13) for system (3.14) at $\lambda = 1$.

(iv) If (A_1, F, b) is a solution of (3.14), then system (6.7) is solvable. Let g be a solution of (6.7), and let X denote the last column of g. Then X is a local isometric immersion of S^n in S^{n+m}.

Next we study submanifolds associated to the $G_{m,n}$- system II (3.10). We will show that each solution of (3.10) gives rise to an $O(n)$-family of submanifolds with common line of curvature coordinates. In order to simplify the notation, we make the following definition:

6.6 Definition. Let $m > n$, $\mathcal{O}$ a domain in R^n, and $X_i : \mathcal{O} \to R^m$ an immersion with flat and non-degenerate normal bundle for $1 \leq i \leq n$. $(X_1, \cdots, X_n)$ is called a *n-tuple in R^m of type $O(n)$* ($O(n-1,1)$ resp.) if

(i) the normal plane of $X_i(x)$ is parallel to the normal plane of $X_j(x)$ for all $1 \leq i, j \leq n$ and $x \in \mathcal{O}$,

(ii) there exists a common parallel normal frame $\{e_\alpha\}_{\alpha=n+1}^m$,

(iii) $x \in \mathcal{O}$ is a spherical (hyperbolic resp.) line of curvature coordinate system (cf. Definition 5.5) with respect to e_α for each X_i such that the fundamental forms for X_j are

$$\begin{aligned} I_j &= \sum_{i=1}^{n} b_{ji}^2 dx_i^2, \\ II_j &= \sum_{i=1,k=1}^{n,m-n} b_{ji} g_{ki} dx_i^2 e_{n+k} \end{aligned} \tag{6.8}$$

for some $O(n)$-valued $((O(n-1,1)$- resp.) map $B=(b_{ij})$ and a $\mathcal{M}_{(m-n)\times n}$-valued map $G=(g_{ij})$.

We note that an n-tuple in R^m of type $O(n)$ or $O(n-1,1)$ is an n-tuple of parametrized n-dimensional submanifolds in R^m and the parametrization is a spherical or hyperbolic line of curvature coordinates.

6.7 Theorem. *Let $(X_1,\cdots,X_n)$ be an n-tuple in R^m of type $O(n)$, $e_{n+1},\cdots,$ e_m common parallel normal frame, and $(x_1,\cdots,x_n)$ a common spherical line of curvature coordinates for all X_j's such that the two fundamental forms I_j, II_j for X_j are given by (6.8). Set $f_{ij}=(b_{1j})_{x_i}/b_{1i}$ if $i\neq j$, $f_{ii}=0$, and $F=(f_{ij})$. If all entries of G are nonzero, then (F,G,B) is a solution of (3.10), the $G_{m,n}$-system II.*

PROOF. It follows from the definition of n-tuples that

$$\omega_1^{(j)} = b_{j1}dx_1,\ \omega_2^{(j)} = b_{j2}dx_2,\ \cdots,\ \omega_n^{(j)} = b_{jn}dx_n$$

is a dual 1-frame for X_j, and $\omega_{i,n+k}^{(j)} = g_{ki}dx_i$ for each X_j. Note that $\omega_{i,n+k}^{(j)}$ is independent of j. By Corollary 5.3, the Levi-Civita connection 1-form for the metric I_j is

$$\omega_{ik}^{(j)} = -f_{ik}^{(j)}dx_k + f_{ki}^{(j)}dx_i, \quad \text{where } f_{ik}^{(j)} = \frac{(b_{jk})_{x_i}}{b_{ji}}.$$

Since

$$d\omega_{i,n+k}^{(j)} = -\sum_{r=1}^{n} \omega_{ir}^{(j)} \wedge \omega_{r,n+k}^{(j)}$$

for $1\le k\le m-n$ and $g_{k1},\cdots,g_{kn}$ are non-zero, Cartan's Lemma and Corollary 5.3 imply that

$$f_{ir}^{(j)} = \frac{(g_{kr})_{x_i}}{g_{ki}},$$

which is independent of j. Hence

$$\omega_{ij}^{(j)} = \omega_{ij}^{(1)} = -f_{ij}dx_j + f_{ji}dx_i.$$

The structure equation, Gauss-Codazzi-Ricci equations for $X_1,\cdots,X_n$ imply that (F,G,B) is a solution of (3.10). ∎

The converse is also true:

6.8 Theorem. *(n-tuples in R^m of type $O(n)$). Suppose $(F,G,B) : R^n \to gl_*(n) \times \mathcal{M}_{(m-n)\times n} \times O(n)$ is a solution of the $G_{m,n}$-system II (3.10). Let*

$$F = (f_{ij}), \quad G = (g_{ij}), \quad B = (b_{ij}).$$

Then:

(i)

$$\tau = \sum_{i=1}^{n} \begin{pmatrix} -FC_i + C_iF^t & C_iG^t \\ -GC_i & 0 \end{pmatrix} dx_i$$

is a flat $o(m)$-valued connection 1-form. Hence there exists $A : R^n \to O(m)$ such that

$$A^{-1}dA = \tau = \sum_{i=1}^{n} \begin{pmatrix} -FC_i + C_iF^t & C_iG^t \\ -GC_i & 0 \end{pmatrix} dx_i. \tag{6.9}$$

(ii) Write $A = (A_1, A_2)$ with $A_1 \in \mathcal{M}_{m\times n}$ and $A_2 \in \mathcal{M}_{m\times(m-n)}$. Then

$$\sum_{i=1}^{n} A_1C_iB^t dx_i$$

is exact. So there exists a map $X : R^n \to \mathcal{M}_{m\times n}$ such that

$$dX = -\sum_{i=1}^{n} A_1C_iB^t dx_i. \tag{6.10}$$

(iii) Suppose all the entries of B are non-zero. Let $X_j : R^n \to R^m$ denote the j-th column of X (solution of (6.10)), and e_i denote the i-th column of A. Then $(X_1, \cdots, X_n)$ is an n-tuple in R^m of type $O(n)$. In fact,

(1) $e_1, \cdots, e_n$ are tangent to X_j for all $1 \le j \le n$; so the tangent planes of $X_1, \cdots, X_n$ are parallel,

(2) $\{e_{n+1}, \cdots, e_m\}$ is a parallel normal frame for each X_j,

(3) the two fundamental forms for the immersion X_j are

$$\begin{aligned} I_j &= \sum_{i=1}^{n} b_{ji}^2 dx_i^2, \\ II_j &= -\sum_{i=1,k=1}^{n,m-n} g_{ki}b_{ji}dx_i^2 e_{n+k}. \end{aligned} \tag{6.11}$$

PROOF. By Proposition 3.7, θ_λ^{II} defined by (3.9) is flat for all $\lambda \in C$. In particular,

$$\theta_0^{II} = \begin{pmatrix} \tau & 0 \\ 0 & 0 \end{pmatrix}$$

is flat.

The gauge transformation of θ_λ^{II} by $\begin{pmatrix} A & 0 \\ 0 & I \end{pmatrix}$ is

$$\begin{pmatrix} A & 0 \\ 0 & I \end{pmatrix} * \theta_\lambda^{II} = \sum_{i=1}^{n} \begin{pmatrix} 0 & -\lambda A_1 C_i B^t \\ \lambda B C_i A_1^t & 0 \end{pmatrix}, \tag{6.12}$$

which is flat for all $\lambda \in C$. It follows from Proposition 2.6 that $\sum_{i=1}^n A_1 C_i B^t dx_i$ is exact. This proves (ii).

Equate the j-th column of equation (6.10) to get

$$dX_j = -\sum_{k=1}^{n} b_{jk} e_k dx_k.$$

So $I_j = \sum_k b_{jk}^2 dx_k^2$. The rest of (iii) follows from the fact that $A^{-1}dA = \tau$. ∎

6.9 Corollary. *Let $(X_1, \cdots, X_n)$ be an n-tuple in R^m of type $O(n)$, and $p \in O(m)$ and $q \in O(n)$ constant matrices. Then $p(X_1, \cdots, X_n)q$ is also an n-tuple in R^m of type $O(n)$.*

PROOF. We call a local orthonormal frame $A = (e_1, \cdots, e_m)$ an adapted frame for the n-tuple $(X_1, \cdots, X_n)$ of type $O(n)$ if $e_1, \cdots, e_n$ are common principal curvature directions and $e_{n+1}, \cdots, e_m$ are a common parallel normal frame for each X_j. By Theorem 6.7, there exist G and B such that (F, G, B) is a solution of system (3.10). A direct computation shows that Xq is an n-tuple with Aq as an adapted frame and the corresponding solution of (3.10) is the same (F, G, B).

Note that if A is a solution of (6.9), then so is pA. By Theorem 6.8 (ii), pX is an n-tuple of type $O(n)$ in R^m. ∎

The immersion X_j in Theorem 6.8 can be obtained either by solving the system (6.10) using integration or by an analogue of Sym's formula (cf. [Sy]) below.

6.10 Proposition. *Suppose $\theta(x, \lambda) = \alpha_1(x)\lambda + \alpha_0(x)$ is a flat $\mathcal{G}$-valued connection 1-form on $x \in R^n$, and $E(x, \lambda)$ is a trivialization of $\theta(x, \lambda)$, i.e., $E^{-1}dE = \theta$. Set*

$$Y(x) = \frac{\partial E}{\partial \lambda}(x, 0)E^{-1}(x, 0).$$

Then $dY = E(x, 0)\alpha_1(x)E(x, 0)^{-1}$.

PROOF. A direct computation gives

$$\begin{aligned} dY &= \left(\frac{\partial}{\partial \lambda}(dE)\right) E^{-1} - \frac{\partial E}{\partial \lambda} E^{-1} dE E^{-1}\Big|_{\lambda=0} \\ &= \left(\frac{\partial}{\partial \lambda}(E\theta)\right) E^{-1}\Big|_{\lambda=0} - \frac{\partial E}{\partial \lambda}(x, 0)\theta(x, 0)E(x, 0)^{-1} \\ &= E(x, 0)\alpha_1(x)E(x, 0)^{-1}. \end{aligned}$$ ∎

6.11 Corollary. *Let $E(x,\lambda)$ be a frame for a solution ξ of the $G_{m,n}$-system (3.1), and*

$$Y(x) = \frac{\partial E}{\partial \lambda}(x,0)E^{-1}(x,0).$$

Then:
(i) $Y = \begin{pmatrix} 0 & X \\ -X^t & 0 \end{pmatrix}$ *for some* $X \in \mathcal{M}_{m\times n}$.
(ii) $X = (X_1, \cdots, X_n)$ *is an n-tuple in R^m of type $O(n)$.*
(iii) $dX = -A_1\delta B^t$, *where* $\delta = \text{diag}(dx_1, \cdots, dx_n)$. *In other words, X satisfies (6.10).*

PROOF. Since E satisfies the reality condition, $E(x,0) \in O(m) \times O(n)$. Write $E(x,0) = \begin{pmatrix} A(x) & 0 \\ 0 & B(x) \end{pmatrix}$ and set

$$\delta = \text{diag}(dx_1, \cdots, dx_n), \quad \beta = \begin{pmatrix} \delta \\ 0 \end{pmatrix}.$$

It follows from Proposition 6.10 and the fact that the Lax connection of (3.1) is $\theta(x,\lambda) = \alpha_0\lambda + \alpha_1$ with

$$\alpha_1 = \begin{pmatrix} 0 & -A_1\delta B^t \\ B\delta A_1^t & 0 \end{pmatrix},$$

that

$$dY = E(x,0)\begin{pmatrix} 0 & -\beta \\ \beta & 0 \end{pmatrix}E(x,0)^{-1} = \begin{pmatrix} 0 & -A_1\delta B^t \\ B\delta A_1^t & 0 \end{pmatrix}.$$

So $dX = -A_1\delta B^t$, i.e., X solves (6.10). ∎

We end this section by studying the $G_{m,2}$-system II (3.10).

6.12 Proposition. *The $G_{m,2}$-system II (3.10) is the Gauss-Codazzi equations for a surface in R^m that admits spherical line of curvature coordinates.*

PROOF. Suppose M is a surface in R^m, which admits a spherical line of curvature coordinate system (x_1, x_2) with respect to a parallel normal frame $e_3, \cdots, e_m$. Then there exist a function u and a $\mathcal{M}_{(m-2)\times 2}$ -valued map $G = (g_{ij})$ such that

$$I = \cos^2 u\, dx_1^2 + \sin^2 u\, dx_2^2,$$
$$II = \sum_{j=1}^{m-2} (g_{j1}\cos u\, dx_1^2 + g_{j2}\sin u\, dx_2^2)e_{2+j}.$$

The Gauss-Codazzi-Ricci equations are the following system for u, g_{j1}, g_{j2}:

$$\begin{cases} u_{x_1x_1} - u_{x_2x_2} + \sum_{i=1}^{m-2} g_{i1}g_{i2} = 0, \\ (g_{i2})_{x_1} = g_{i1}u_{x_1}, \\ (g_{i1})_{x_2} = -g_{i2}u_{x_2}. \end{cases} \tag{6.13}$$

Let

$$B = \begin{pmatrix} \cos u & \sin u \\ -\sin u & \cos u \end{pmatrix}, \quad F = \begin{pmatrix} 0 & u_{x_1} \\ -u_{x_2} & 0 \end{pmatrix}, \quad G = (g_{ij}). \tag{6.14}$$

Then (F, G, B) is a solution of the $G_{m,2}$ system II (3.10).

Conversely, if (F, G, B) is a solution of the $G_{m,2}$-system II (3.10), then we may assume $B = \begin{pmatrix} \cos u & \sin u \\ -\sin u & \cos u \end{pmatrix}$. If $\sin u \cos u \neq 0$, then by the third equation of (3.10), we have $f_{12} = u_{x_1}$ and $f_{21} = -u_{x_2}$, i.e., $F = \begin{pmatrix} 0 & u_{x_1} \\ -u_{x_2} & 0 \end{pmatrix}$. Let g_{ij} denote the (i, j)-th entry of G. Then system (3.10) is system (6.13). ∎

6.13 Corollary. *Let $X_1 : \mathcal{O} \to R^m$ be an immersion and $(x, y) \in \mathcal{O}$ the spherical line of curvature coordinates with respect to a parallel normal frame $e_3, \cdots, e_m$. Then there exists an immersion surface X_2 unique up to translation such that (X_1, X_2) is a 2-tuple in R^m of type $O(2)$. Moreover, the fundamental forms of X_1, X_2 are respectively given by*

$$\begin{aligned} I_1 &= \cos^2 u\, dx_1^2 + \sin^2 u\, dx_2^2, \quad II_1 = \sum_{j=1}^{m} (g_{j1} \cos u\, dx_1^2 + g_{j2} \sin u\, dx_2^2) e_{2+j}, \\ I_2 &= \sin^2 u\, dx_1^2 + \cos^2 u\, dx_2^2, \quad II_2 = \sum_{j=1}^{m-2} (-g_{j1} \sin u\, dx_1^2 + g_{j2} \cos u\, dx_2^2) e_{2+j}. \end{aligned} \tag{6.15}$$

It follows from the Gauss equation (5.3) that

$$K_1 = -K_2 = \frac{\sum_{j=1}^{m-2} g_{j1} g_{j2}}{\sin u \cos u}.$$

So we have

6.14 Corollary. *Let $\mathcal{O}$ be an open subset of R^2, $(X_1, X_2) : \mathcal{O} \to R^m \times R^m$ a 2-tuple in R^m of type $O(2)$, and $(x_1, x_2) \in \mathcal{O}$ spherical line of curvature coordinates. Then*

$$K_2(x) = -K_1(x),$$

where K_1, K_2 are respectively the Gaussian curvature of X_1, X_2.

6.15 Definition. If (M_1, M_2) is a 2-tuple in R^m of type $O(2)$ ($O(1,1)$ resp.) with a parallel normal frame $e_3, \cdots, e_m$ and spherical (hyperbolic resp.) line of curvature coordinates, then we call M_2 a *C-dual* of M_1. Note that any two C-duals of M_1 are differed by a translation.

6.16 Example. Recall that given a surface M in R^3 with curvature -1, there exist spherical line of curvature coordinates x_1, x_2 such that the two fundamental forms are

$$\begin{aligned} I &= \cos^2 u\, dx_1^2 + \sin^2 u\, dx_2^2, \\ II &= \sin u \cos u\, (dx_1^2 - dx_2^2), \end{aligned}$$

and u satisfies

$$u_{x_1x_1} - u_{x_2x_2} = \sin u \cos u. \tag{SGE}$$

This implies that $(u, \sin u, -\cos u)$ is a solution of (6.13). Let $X(x_1, x_2)$ denote the immersion of M. Then (X, e_3) is a 2-tuple in R^3 of type $O(2)$, where e_3 is the unit normal of M, which is an open subset of S^2.

7. Submanifolds associated to $G^1_{m,n}$- systems

In this section, we describe submanifolds whose fundamental equations are given by $G^1_{m,n}$-systems.

Let $R^{k,1}$ denote the Lorentz space equipped with the non-degenerate bilinear form of index one:

$$\langle x, y\rangle_1 = x_1y_1 + \ldots + x_ky_k - x_{k+1}y_{k+1}.$$

The moving frame computation for submanifolds in $R^{k,1}$ can be carried out in a similar way as for submanifolds in R^{k+1} except that the Levi-Civita connection 1-form (ω_{ij}) of the flat Lorentzian metric $\langle , \rangle_1$ is $o(k,1)$-valued.

Let H^k denote the k-dimensional simply connected space form of sectional curvature -1 (i.e., a *hyperbolic k-space*). It is well-known that

$$\{x \in R^{k,1} \mid \langle x, x\rangle_1 = -1, x_{k+1} > 0\}$$

with the induced metric is isometric to H^k. We need the following Proposition later, which can be proved using a direct computation (cf. Chap. 2 of [PT]).

7.1 Proposition. *Let $v_0 \in R^{k,1}$ be a constant non-zero vector, $c \in R$ a constant, and $N_{v,c}$ the hypersurface defined by*

$$N_{v_0,c} = \{x \in H^k \mid \langle x, v_0\rangle_1 = c\}.$$

Then $N_{v,c}$ is a totally umbilic hypersurface of H^k with constant sectional curvature

$$\frac{-\langle v_0, v_0\rangle_1}{c^2 + \langle v_0, v_0\rangle_1}.$$

We use methods similar to those in section 6 to find submanifolds whose fundamental equations are the various $G^1_{m,n}$-systems. For the $G^1_{m,n}$-system I, let (θ_{ij}) denote the connection 1-form θ^I_λ defined by (4.3) at $\lambda = 1$. Then (θ_{ij}) is a $o(m+n,1)$-valued 1-form and

$$\begin{cases} \theta_{m+i,m+j} = -\epsilon_i\epsilon_j f_{ji} dx_j + f_{ij} dx_i, & 1 \le i,j \le n+1, \\ \theta_{m+i,\alpha} = a_{\alpha i} dx_i, & 1 \le i \le n+1, 1 \le \alpha \le m, \\ \theta_{\alpha\beta} = 0, & 1 \le \alpha, \beta \le m, \end{cases}$$

where $\epsilon_1 = \cdots = \epsilon_n = -\epsilon_{n+1} = 1$. Let

$$\begin{cases} \omega_{ij} = \theta_{m+i,m+j}, & 1 \le i,j \le n+1, \\ \omega_{i\alpha} = \theta_{m+i,\alpha}, & 1 \le i \le n+1, 1 \le \alpha \le m, \\ \omega_{\alpha\beta} = \theta_{\alpha\beta}, & 1 \le \alpha, \beta \le m. \end{cases}$$

Then the flatness of (θ_{ij}) is exactly the Gauss-Codazzi-Ricci equations for $(n+1)$-dimensional flat, Lorentzian submanifolds in $R^{n+m,1}$. So we get

7.2 Theorem. *(Local isometric immersions of $R^{n,1}$ in $R^{m+n,1}$). Let X be a local isometric immersion of $R^{n,1}$ in $R^{n+m,1}$ with flat and non-degenerate normal bundle. Fix a local parallel normal frame $e_{n+2}, \cdots, e_{n+m+1}$. Then:*

(i) There exist a line of curvature coordinates $(x_1, ..., x_{n+1})$ and a $\mathcal{M}_{m\times(n+1)}$-valued map $A_1 = (a_{ij})$ and a map $b = (b_1, ..., b_{n+1})$ such that $A_1^t A_1 = I$ and the first and second fundamental forms of X are given by

$$I = \sum_{i=1}^{n+1} \epsilon_i b_i^2 dx_i^2, \quad II = \sum_{i=1}^{n+1} \sum_{j=2}^{m-n} b_i a_{ji} dx_i^2 e_{n+j}, \tag{7.1}$$

where $\epsilon_1 = \cdots = \epsilon_n = -\epsilon_{n+1} = 1$.

(ii) Let $f_{ij} = (b_i)_{x_j}/b_j$ for $i \neq j$, $f_{ii} = 0$, and $F = (f_{ij})$. Then (A_1, F) is a solution of the $G^1_{m,n}$-system I (4.4).

(iii) Conversely, if (A_1, F) is a solution of (4.4) and $b_1, \cdots, b_{n+1}$ satisfies $(b_i)_{x_j} = f_{ij} b_j$ for all $i \neq j$, then there exists a local isometric immersion of $R^{n,1}$ in $R^{m+n,1}$ such that the two fundamental forms are given by (7.1), where $A_1 = (a_{ij})$ and $F = (f_{ij})$.

Next we consider the partial $G^1_{m,n}$-system I of n variables, system (4.10). Let θ_{ij} denote the (i,j)-th entry of τ^I_λ, defined by (4.9), at $\lambda = 1$. So we have

$$\begin{cases} \theta_{\alpha\beta} = 0, & 1 \leq \alpha, \beta \leq m, \\ \theta_{m+i,m+j} = -f_{ji} dx_j + f_{ij} dx_i, & 1 \leq i, j \leq n, \\ \theta_{m+i,\alpha} = a_{\alpha i} dx_i, & 1 \leq i \leq n, 1 \leq \alpha \leq m, \\ \theta_{m+i,m+n+1} = b_i dx_i, & 1 \leq i \leq n. \end{cases}$$

Let

$$\begin{cases} \omega_i = \theta_{m+i,m+n+1}, & 1 \leq i \leq n, \\ \omega_{ij} = \theta_{m+i,m+j}, & \text{if } 1 \leq i, j \leq n, \\ \omega_{i,n+j} = \theta_{m+i,j}, & \text{if } 1 \leq i \leq n, \quad 1 \leq j \leq m, \\ \omega_{n+i,n+j} = \theta_{ij}, & \text{if } 1 \leq i, j \leq m. \end{cases}$$

Then system (4.10) (given by the flatness of τ^I_1) is exactly the Gauss-Codazzi-Ricci equations for local isometric immersions of H^n in $H^{n+m} \subset R^{n+m,1}$ with flat and non-degenerate normal bundle. We summarize this below.

7.3 Theorem. *(Local isometric immersion of H^n in H^{n+m}). Let X be a local isometric immersion of H^n in H^{n+m} with flat and non-degenerate normal bundle, and $\{e_\alpha\}$ a parallel normal frame. Then:*

(i) There exist a line of curvature coordinate system $(x_1, \ldots, x_n)$, a $\mathcal{M}_{m\times n}$-valued map $A_1 = (a_{ij})$ satisfying $A_1^t A_1 = I$, and a $\mathcal{M}_{n\times 1}$-valued map $b = (b_1, \cdots, b_n)^t$ such that the two fundamental forms of X are given by

$$I = \sum_i b_i^2 \, dx_i^2, \quad II = \sum_{i=1,j=1}^{n,m-n} a_{ji} b_i dx_i^2 e_{n+j}. \tag{7.2}$$

(ii) Let $f_{ij} = (b_i)_{x_j}/b_j$ if $i \neq j$, $f_{ii} = 0$ for all $1 \leq i \leq n$, and $F = (f_{ij})$. Then (A_1, F, b) is a solution of the partial $G^1_{m,n}$-system (4.10).

(iii) Let $e_i = \frac{1}{b_i} X_{x_i}$, *and* $g = (e_{n+1}, \cdots, e_{n+m}, e_1, \cdots, e_n, X)$. *Then*

$$g^{-1}dg = \sum_i \begin{pmatrix} 0 & -A_1 C_i & 0 \\ C_i A_1^t & -F^t C_i + C_i F & C_i b \\ 0 & b^t C_i & 0 \end{pmatrix} dx_i, \tag{7.3}$$

which is equal to the Lax connection τ_λ^I *defined in (4.9) for system (4.10) at* $\lambda = 1$.

(iv) If (A_1, F, b) *is a solution of (4.10), then system (7.3) is solvable.*

(v) Let g *be a solution of (7.3). Then the last column of* g *is an isometric immersion of a constant sectional curvature* -1 *submanifold of* H^{n+m}.

(vi) Let M *be the* n *dimensional submanifold given in (v). Then* M^n *lies in a totally umbilic hypersurface of* H^{n+m} *if and only if there is a constant vector* $w \in \mathcal{M}_{m\times 1}$ *such that* $b = A_1^t w$. *Moreover,* M^n *lies in a flat totally umbilic hypersurface if and only if* $\|w\| = 1$.

PROOF. Statements (i)-(v) follow from standard submanifold theory. To prove (vi), note that every umbilic hypersurface of H^{n+m} is the intersection of H^{n+m} with a hyperplane (cf. [PT]), i.e., it is of the form

$$N_{v,c} = \{x \in H^{n+m} \mid \langle x, v\rangle_1 = c\}$$

for some constant $v \in R^{n+m,1}$ and $c \in R$. By Proposition 7.1, $N_{v,c}$ has constant sectional curvature

$$-\frac{\langle v, v\rangle_1}{c^2 + \langle v, v\rangle_1}.$$

Suppose the image of X lies in $N_{v,c}$. Then $< dX, v >_1 = 0$, which implies that v is normal to X. But v is constant, so v is a parallel normal vector field of M as a submanifold of $R^{n+m,1}$. Hence

$$v = \sum_{k=1}^m v_k e_{n+k} + v_{m+1} X \tag{7.4}$$

for some constants $v_1, \cdots, v_{m+1}$. It follows from $\langle X, v\rangle_1 = c$ and $\langle X, X\rangle_1 = -1$ that $v_{m+1} = -c$. Differentiate (7.4) to get

$$c\omega_i = cb_i dx_i = \sum_{k=1}^m v_k \omega_{i,n+k} = \sum_{k=1}^m v_k a_{ki} dx_i,$$

so

$$cb_i = \sum v_k a_{ki}. \tag{7.5}$$

Let $\hat{v} = (v_1, \cdots, v_m)^t$, and $w = \hat{v}/c$. Then (7.5) implies that $b = A_1^t w$. Since $\langle v, v\rangle_1 = \|\hat{v}\|^2 - c^2$, it follows from the formula for the sectional curvature that $N_{v,c}$ is flat if and only if $\langle v, v\rangle_1 = 0$, i.e., $\|\hat{v}\|^2 = c^2$, or equivalently, $\|w\| = 1$. ∎

For $m = n$, the above theorem was proved by Terng in [Te2].

7.4 Theorem. *($(n+1)$-tuples in R^m of type $O(n,1)$). Suppose $m > n+1$, and*

$$(F, G, B) : R^{n+1} \to gl_*(n+1) \times \mathcal{M}_{(m-n-1)\times(n+1)} \times O(n,1)$$

is a solution of the $G^1_{m,n}$-system II (4.6). Let $F = (f_{ij})$, $G = (g_{ij})$, and $B = (b_{ij})$. Then:

(i)

$$\omega = \sum_{i=1}^{n+1} \begin{pmatrix} -FC_i + C_iF^t & C_iG^t \\ -GC_i & 0 \end{pmatrix} dx_i \tag{7.6}$$

is a flat $o(m)$-valued connection 1-form. Hence there exists $A : R^{n+1} \to O(m)$ such that

$$A^{-1}dA = \omega = \sum_{i=1}^{n+1} \begin{pmatrix} -FC_i + C_iF^t & C_iG^t \\ -GC_i & 0 \end{pmatrix} dx_i. \tag{7.7}$$

(ii) Let $A = (A_1, A_2)$ with $A_1 \in \mathcal{M}_{m\times(n+1)}$. Then $\sum_{i=1}^{n+1} A_1C_iJB^{-1}dx_i$ is exact, so there exists a map $X : R^{n+1} \to \mathcal{M}_{m\times(n+1)}$ such that

$$dX = -\sum_{i=1}^{n+1} A_1C_iB^tJdx_i. \tag{7.8}$$

(iii) Suppose all the entries of the j-th row of B are non-zero. Let $X_j : R^{n+1} \to R^m$ denote the j-th column of X, and e_i denote the i-th column of A. Then $(X_1, \cdots, X_{n+1})$ is an $(n+1)$-tuple in R^m of type $O(n,1)$, and the two fundamental forms for the immersion X_j are

$$\begin{aligned} I_j &= \sum_{i=1}^{n+1} b_{ji}^2 dx_i^2, \\ II_j &= -\sum_{i=1,k=1}^{n+1,m-n-1} g_{ki}b_{ji}dx_i^2 e_{n+k+1}. \end{aligned} \tag{7.9}$$

(iv) Conversely, suppose $(X_1, \cdots, X_{n+1})$ is an $(n+1)$-tuple in R^m of type $O(n,1)$ such that the two fundamental forms are of the form (7.9) with respect to a common parallel normal frame $e_{n+2}, \cdots, e_m$ for some $B = (b_{ij})$ and $G = (g_{ij})$. Then (F, G, B) is a solution of (4.6), where

$$F = (f_{ij}), \quad f_{ij} = \frac{(b_{1j})_{x_i}}{b_{1i}}.$$

PROOF. The proof is similar to that of Theorem 6.8. We only give the proof of (iii) here. Let ω_{ij} denote the (i,j)-th entry of ω (defined by (7.6)), i.e.,

$$\omega_{ij} = \begin{cases} -f_{ij}dx_j + f_{ji}dx_i, & 1 \le i,j \le n+1, \\ g_{j-n-1,i}dx_i, & 1 \le i \le n+1 \text{ and } n+2 \le j \le m, \\ 0, & n+2 \le i,j \le m. \end{cases} \tag{7.10}$$

Let e_i denote the i-th column of A (defined by (7.7)), i.e., $A = (e_1, \cdots, e_m)$. Since $dA = A\omega$, we have

$$de_i = \sum_{j=1}^{m} \omega_{ji} e_j, \quad 1 \le i \le m. \tag{7.11}$$

By Theorem 7.4 (ii),

$$dX_r = -\sum_{i=1}^{n+1} b_{ri}\epsilon_r e_i dx_i, \tag{7.12}$$

where X_r is the r-th column of X. So $\{e_1, \cdots, e_{n+1}\}$ is a common orthonormal tangent frame, and

$$\omega_1^{(r)} = -\epsilon_r b_{r1} dx_1, \cdots, \omega_{n+1}^{(r)} = -\epsilon_r b_{r,n+1} dx_{n+1}$$

is the dual coframe for X_r. Since

$$\omega_{i,n+1+j}^{(r)} = \langle de_{n+1+j}, e_i \rangle_1 = \omega_{i,n+j+1} = g_{ji} dx_i,$$

$\{e_{n+2}, \cdots, e_m\}$ is a common parallel normal frame and $(x_1, \cdots, x_{n+1})$ is a hyperbolic line of curvature coordinate system for each X_r. ∎

7.5 Corollary. *Let $(X_1, \cdots, X_{n+1})$ be an $(n+1)$-tuple in R^m of type $O(n,1)$, and $p \in O(m)$, $q \in O(n,1)$ constant matrices. Then $p(X_1, \cdots, X_{n+1})q$ is also a $(n+1)$-tuple in R^m of type $O(n,1)$.*

7.6 Corollary. *Let v be a solution of the $G^1_{m,n}$-system (4.2), E a frame for v, and $Y = (\frac{\partial E}{\partial \lambda} E^{-1})(x, 0)$. Then:*

(i) $Y = \begin{pmatrix} 0 & X \\ -JX^t & 0 \end{pmatrix}$ for some $\mathcal{M}_{m\times(n+1)}$-valued map X and $J = I_{n,1} = \mathrm{diag}(1, \cdots, 1, -1)$.

(ii) $X = (X_1, \cdots, X_n)$ is an $(n+1)$-tuple in R^m of type $O(n,1)$.

8. $G^1_{m,1}$-systems and isothermic surfaces

We study the relation between the $G^1_{m,1}$-system and isothermic surfaces in R^m.

8.1 Proposition. *The $G^1_{m,1}$-system II (4.6) is the Gauss-Codazzi-Ricci equations for surfaces in R^m admitting hyperbolic line of curvature coordinates.*

PROOF. Let $\mathcal{O}$ be a domain in R^2, $X : \mathcal{O} \to R^m$ an immersion with flat and non-degenerate normal bundle, and $(x_1, x_2) \in \mathcal{O}$ a hyperbolic line of curvature coordinate system with respect to the parallel normal frame $\{e_3, \cdots, e_m\}$. Then there exist u and g_{ki} for $1 \leq k \leq m-2$ and $i = 1, 2$ such that the two fundamental forms are

$$\begin{aligned} I &= \cosh^2 u\, dx_1^2 + \sinh^2 u\, dx_2^2, \\ II &= \sum_{k=1}^{m-2} (g_{k1} \cosh u\, dx_1^2 + g_{k2} \sinh u\, dx_2^2) e_{k+2}. \end{aligned} \tag{8.1}$$

The Gauss-Codazzi-Ricci equations for X are

$$\begin{cases} u_{x_1 x_1} + u_{x_2 x_2} + \sum_{k=1}^{m-2} g_{k1} g_{k2} = 0, \\ (g_{k1})_{x_2} = u_{x_2} g_{k2}, \\ (g_{k2})_{x_1} = u_{x_1} g_{k1}. \end{cases} \tag{8.2}$$

This is exactly the $G^1_{m,1}$-system (4.6) for (F, G, B), where

$$F = \begin{pmatrix} 0 & u_{x_1} \\ u_{x_2} & 0 \end{pmatrix}, \quad G = (g_{ki}), \quad B = \begin{pmatrix} \cosh u & \sinh u \\ \sinh u & \cosh u \end{pmatrix} \in O(1,1). \tag{8.3}$$

Conversely, if (F, G, B) is a solution of the $G^1_{m,1}$-system (4.6), then since $B \in O(1,1)$ we may assume

$$B = \begin{pmatrix} \cosh u & \sinh u \\ \sinh u & \cosh u \end{pmatrix}.$$

The third equation of (4.6) implies that $f_{12} = u_{x_1}$ and $f_{21} = u_{x_2}$, i.e., (F, G, B) is of the form (8.3). Write equation (4.6) for (F, G, B) in terms of u and g_{ki} to get equation (8.2). This completes the proof. ∎

8.2 Corollary. *Let $\mathcal{O}$ be a domain of R^2, and $X_1 : \mathcal{O} \to R^m$ an immersion with flat normal bundle and $(x, y) \in \mathcal{O}$ a hyperbolic line of curvature coordinate system with respect to a parallel normal frame $e_3, \cdots, e_m$. Then there exists an immersion X_2 unique up to translation such that (X_1, X_2) is a 2-tuple in R^m of type $O(1,1)$. Moreover, the fundamental forms of X_1, X_2 are given respectively by*

$$\begin{aligned} &\begin{cases} I_1 = \cosh^2 u\, dx_1^2 + \sinh^2 u\, dx_2^2, \\ II_1 = \sum_{j=1}^{m-2} (g_{j1} \cosh u\, dx_1^2 + g_{j2} \sinh u\, dx_2^2) e_{2+j}, \end{cases} \\ &\begin{cases} I_2 = \sinh^2 u\, dx_1^2 + \cosh^2 u\, dx_2^2, \\ II_2 = \sum_{j=1}^{m-2} (-g_{j1} \sinh u\, dx_1^2 - g_{j2} \cosh u\, dx_2^2) e_{2+j}. \end{cases} \end{aligned} \tag{8.4}$$

Let K_1, K_2 denote the Gaussian curvature of X_1 and X_2 respectively. The Gauss equation implies that

$$K_1 = K_2 = \frac{\sum_{k=1}^{m-2} g_{k1}g_{k2}}{\sinh u \cosh u}.$$

Hence we get

8.3 Corollary. *Let $(X_1, X_2) : \mathcal{O} \to R^m$ be a 2-tuple in R^m of type $O(1,1)$ such that $(x_1, x_2) \in \mathcal{O}$ is a hyperbolic line of curvature coordinate system. Then*

$$K_2(x) = K_1(x)$$

where K_1, K_2 are respectively the Gaussian curvature of X_1, X_2.

Next we describe the relation between 2-tuple in R^m of type $O(1,1)$ and isothermic surfaces in R^m. The following definition, given by Burstall in [Bu], generalizes the classical notion of isothermic surfaces in R^3 ([Da]).

8.4 Definition. Let $\mathcal{O}$ be a domain in R^2. An immersion $X : \mathcal{O} \to R^m$ is called *isothermic* if it has flat normal bundle and the two fundamental forms are of the form

$$I = e^{2u}(dx_1^2 + dx_2^2), \quad II = \sum_{k=1}^{m-2} e^u(g_{k1}dx_1^2 + g_{k2}dx_2^2)e_{2+k}$$

with respect to some parallel normal frame $e_3, \cdots, e_m$, or equivalent $(x_1, x_2) \in \mathcal{O}$ is conformal and line of curvature coordinate system for X.

A direct computation gives

8.5 Proposition. *The Gauss-Codazzi-Ricci equation for isothermic surfaces in R^m is (8.2).*

It was first noted by Cieśliński, Goldstein and Sym in [CGS] that equation for isothermic surfaces in R^3 has a Lax pair. There have been many papers ([Ci], [BHPP], [Bu], [HP], [HMN]) using techniques from solition theory to study isothermic surfaces in 3-space.

Next we give a simple relation between isothermic surfaces in R^m and 2-tuples in R^m of type $O(1,1)$.

8.6 Proposition. *Suppose (X_1, X_2) is a 2-tuple in R^m of type $O(1,1)$. Let $Y_1 = X_1 - X_2$ and $Y_2 = X_1 + X_2$. Then both Y_1 and Y_2 are isothermic.*

PROOF. Let $(u, (g_{ij}))$ be the solution of (8.2) associated to (X_1, X_2). We use the same notations as in Theorem 7.4. Write $X = (X_1, X_2)$. By (7.12), we get

$$\begin{cases} dX_1 = -\cosh u\, dx_1 e_1 - \sinh u\, dx_2 e_2, \\ dX_2 = \sinh u\, dx_1 e_1 + \cosh u\, dx_2 e_2. \end{cases}$$

Note that

$$F = \begin{pmatrix} 0 & u_{x_1} \\ u_{x_2} & 0 \end{pmatrix}, \quad \omega_{12} = u_{x_2}dx_1 - u_{x_1}dx_2, \quad \omega_{i,2+j} = g_{ji}dx_i.$$

Use (7.11) to get

$$\begin{cases} \nabla e_1 = \omega_{21} e_2 = (u_{x_2} dx_1 - u_{x_1} dx_2) e_2, \\ \nabla e_2 = \omega_{12} e_1 = -(u_{x_2} dx_1 - u_{x_1} dx_2) e_1, \\ de_{k+2} = \omega_{1,k+2} e_1 + \omega_{2,k+2} e_2 = g_{k1} dx_1 e_1 + g_{k2} dx_2 e_2. \end{cases}$$

Compute directly to get

$$dY_1 = dX_1 - dX_2 = -(\cosh u + \sinh u)(dx_1 e_1 + dx_2 e_2).$$

So the induced metric for Y_1 is $e^{2u}(dx_1^2 + dx_2^2)$. Similarly, we can get explicit formula for dY_2 and de_k for $3 \le k \le m$. This computation implies that the two fundamental forms for Y_1 and Y_2 are

$$\begin{cases} I_1 = e^{2u}(dx_1^2 + dx_2^2), \\ II_1 = -e^u \sum_{k=1}^{m-2} (g_{k1} dx_1^2 + g_{k2} dx_2^2) e_{2+k}, \end{cases} \tag{8.5}$$

$$\begin{cases} I_2 = e^{-2u}(dx_1^2 + dx_2^2), \\ II_2 = e^{-u} \sum_{k=1}^{m-2} (-g_{k1} dx_1^2 + g_{k2} dx_2^2) e_{2+k}. \end{cases} \tag{8.6}$$

Hence Y_1 and Y_2 are isothermic surfaces. ∎

We call (Y_1, Y_2) an *isothermic pair*. If (u, g_{k1}, g_{k2}) is the solution of (8.2) corresponding to Y_1, then $(-u, g_{k1}, -g_{k2})$ is the solution of (8.2) corresponding to Y_2. For $m = 3$, the transformation from Y_1 to Y_2 is the classical Christoffel transformation. Burstall generalizes this result to arbitrary m in [Bu]. In fact, he developed in [Bu] a beautiful theory of isothermic surfaces in R^m and explained its relation to conformal geometry, Clifford algebra and the curved flat system associated to $G^1_{m,1}$. His result motivated us to study the $G_{m,n}$-system II and $G^1_{m,n}$-system II for general m, n.

Next we describe the $o(m+1,1)$ model used by Burstall, Jeromin, Pedit and Pinkall [BHPP] for $m = 3$ and by Burstall [Bu] for general $m \ge 3$. Let

$$\sigma_{k,1} = \begin{pmatrix} I_{k-1} & 0 \\ 0 & \sigma_{1,1} \end{pmatrix}, \qquad r_{k,1} = \begin{pmatrix} I_{k-1} & 0 \\ 0 & r_{1,1} \end{pmatrix},$$
$$\text{where } \sigma_{1,1} = \begin{pmatrix} 0 & 1 \\ 1 & 0 \end{pmatrix}, \quad r_{1,1} = \frac{1}{\sqrt{2}} \begin{pmatrix} 1 & -1 \\ 1 & 1 \end{pmatrix}.$$

Note that the bilinear form defined by $\sigma_{k,1}$ and $I_{k,1}$ are isometric, and

$$r^t_{k,1} \sigma_{k,1} r_{k,1} = I_{k,1}.$$

Let

$$O_\sigma(k,1) = \{\xi \in GL(k+1) \,|\, \xi^t \sigma \xi = \sigma\}.$$

Then $\phi_{k,1}(g) = r^t_{k,1} g r_{k,1}$ is a group isomorphism from $O(k,1)$ to $O_\sigma(k,1)$. If we replace θ^{II}_λ by $\Omega^{II}_\lambda = (\phi_{m+1,1})_*(\theta^{II}_\lambda)$ in Theorem 7.4, then the integration in Theorem 7.4 (ii) gives an isothermic pair $Y = (Y_1, Y_2)$.

The following proposition, which states that a CMC surface in R^3 without umbilic points admits isothermic coordinates, is well-known (for a proof see [PT]).

8.7 Proposition. *Let M be an immersed surface without umbilic points in space form $N^3(c)$, and k_1, k_2 the principal curvature functions. If M has constant mean curvature H and $k_1 > k_2$, then there exists a conformal line of curvature coordinates x, y such that*

$$I = \frac{2}{k_1 - k_2}\left(dx^2 + dy^2\right), \quad II = \left(1 + \frac{H}{k_1 - k_2}\right) dx^2 + \left(-1 + \frac{H}{k_1 - k_2}\right) dy^2.$$

Moreover, if we write $e^{2u} = \frac{2}{k_1 - k_2}$, then the Gauss-Codazzi equation for M is

$$u_{xx} + u_{yy} = e^{-2u} - \left(\frac{H^2}{4} + c\right) e^{2u}. \tag{8.7}$$

8.8 Corollary. *With the same notation as in Proposition 8.7. If $c = -H^2/4$, then equation (8.7) becomes*

$$u_{xx} + u_{yy} = e^{-2u}. \tag{8.8}$$

Moreover, $(u, e^{-u}, -e^{-u})$ is a solution of (8.2) if and only if u is a solution of (8.8).

Proposition 8.7 has a natural generalization to surfaces in R^m observed by Burstall [Bu]. First we recall a definition due to Chen [C]:

8.9 Definition. A surface M in R^m with flat normal bundle is called a *generalized H-surface* if there exists a unit parallel normal field v such that $\langle H, v\rangle =$ constant, where H is the mean curvature vector field.

Note that a generalized H-surface in R^3 is a CMC surface.

The following analogue of Proposition 8.7 can be proved exactly by the same way for generalized H-surfaces in R^m.

8.10 Proposition ([Bu]). *Let M be a generalized H-surface in R^m, H the mean curvature vector, and $(e_3, \cdots, e_m)$ a parallel normal field such that $\langle H, e_3\rangle = c$ is a constant. If the shape operator A_{e_3} has two distinct eigenvalues $k_1 > k_2$, then there exists isothermic coordinates x, y on M such that*

$$I = \frac{2}{k_1 - k_2}\left(dx^2 + dy^2\right),$$
$$II_3 = \langle II, e_3\rangle = \left(1 + \frac{c}{k_1 - k_2}\right) dx^2 + \left(-1 + \frac{c}{k_1 - k_2}\right) dy^2,$$
$$II_\alpha = \langle II, e_\alpha\rangle = g_{\alpha 1} dx^2 + g_{\alpha 2} dy^2, \quad 3 < \alpha \leq m$$

for some $g_{\alpha i}$.

We call the isothermic coordinates x, y in the above Proposition *the canonical isothermic coordinates* for the generalized H-surface M.

The following Proposition generalizes the classical Bonnet transformation for CMC surfaces in R^3. The proof follows from a direct computation as in the classical case.

8.11 Proposition ([Bu]). *Let $\mathcal{O}$ be a domain in R^2, $Y : \mathcal{O} \to R^m$ an isothermic immersion of a generalized H-surface in R^m, $(x_1, x_2) \in \mathcal{O}$ the canonical isothermic coordinates, and e_3 the parallel unit normal field such that $\langle H, e_3 \rangle = 2c$ is a constant. Then:*

(i) If $c \neq 0$, then $(Y, Y - \frac{1}{c} e_3)$ is an isothermic pair.

(ii) If $c = 0$, then (Y, e_3) is an isothermic pair.

Next we study the class of solutions of the $G^1_{3,1}$-system II corresponding to isothermic immersions of minimal surface in R^3. If Y is an isothermic immersion of a minimal surface in R^3 such that

$$I = e^{2u}(dx^2 + dy^2), \quad II = dx^2 - dy^2,$$

then (Y, e_3) is an isothermic pair, where e_3 is the unit normal of Y. It follows from Propositions 8.1, 8.11 and Corollary 8.2 that $((e_3 - Y)/2, (e_3 + Y)/2)$ is a 2-tuple in R^3 of type $O(1,1)$, and the corresponding solution (F, G, B) of the $G^1_{3,1}$-system II (4.6) is given by

$$F = \begin{pmatrix} 0 & u_x \\ u_y & 0 \end{pmatrix}, \quad G = (e^{-u}, -e^{-u}), \quad B = \begin{pmatrix} \cosh u & \sinh u \\ \sinh u & \cosh u \end{pmatrix}. \tag{8.9}$$

The converse is a consequence of the definition of $G^1_{3,1}$-system II, Corollary 7.6, Proposition 8.6 and Corollary (8.8):

8.12 Proposition. *If u is a solution of (8.8), then*

$$v = \begin{pmatrix} 0 & u_x \\ u_y & 0 \\ e^{-u} & -e^{-u} \end{pmatrix}$$

is a solution of the $G^1_{3,1}$-system (4.2). Moreover, let E be a frame for v, i.e.,

$$E^{-1}dE = \theta_\lambda = \begin{pmatrix} \delta v^t - v\delta^t & -\delta J \lambda \\ \delta^t \lambda & -J v^t \delta J + \delta^t v \end{pmatrix},$$

where $\delta = \begin{pmatrix} dx_1 & 0 \\ 0 & dx_2 \\ 0 & 0 \end{pmatrix}$ and $J = \mathrm{diag}(1, -1)$. *Then*

(i) $(\frac{\partial E}{\partial \lambda} E^{-1})(x, y, 0) = \begin{pmatrix} 0 & X(x,y) \\ -JX^t(x,y) & 0 \end{pmatrix}$ *for some $\mathcal{M}_{3\times 2}$-valued map X,*

(ii) $X = (X_1, X_2)$ is a 2-tuple in R^3 of type $O(1,1)$,

(iii) let $Y_1 = X_1 - X_2$ and $Y_2 = X_1 + X_2$, then (Y_1, Y_2) is an isothermic pair such that Y_1 is minimal and Y_2 is the unit normal of Y_1,

(iv) (x, y) is the canonical isothermic coordinate system for minimal surface Y_1.

It follows from Corollary 8.8 that a solution of (8.8) gives a minimal surface in R^3 and a surface in $N^3(-1) = H^3$ with mean curvature 2 unique up to ambient isometries. Proposition 8.12 gives a construction of the minimal immersion in R^3 from the frame of the $G^1_{3,1}$-system. Next we show that the immersion of the corresponding CMC surface in H^3 can also be read easily from the frame of the $G^1_{3,1}$-system II. First we recall that if $E^0(x,y,\lambda)$ is a frame of the solution $v = \begin{pmatrix} F \\ G \end{pmatrix}$ of the $G^1_{3,1}$-system, then $E^0(x,y,0) = \begin{pmatrix} A(x,y) & 0 \\ 0 & B(x,y) \end{pmatrix}$ for some A, B and

$$E(x,y,\lambda) = E^0(x,y,\lambda)\begin{pmatrix} I & 0 \\ 0 & B^{-1}(x,y) \end{pmatrix}$$

is a frame for the solution (F,G,B) of the $G^1_{3,1}$-system II.

8.13 Proposition. *Let u be a solution of (8.8), and (F,G,B) defined by (8.9). Then (F,G,B) is a solution of the $G^1_{3,1}$-system II (4.6). Moreover, let $E(x,y,\lambda)$ be a trivialization of the Lax pair θ^{II}_λ defined by (4.5) for (F,G,B), i.e., $E^{-1}dE = \theta^{II}_\lambda$, and $e_i(x,y)$ the i-th column of $E(x,y,\sqrt{2})$. Then:*
(i) $v_0 = \sqrt{2}\, e_3 + e_4 + e_5$ is independent of (x,y) and $\langle v_0, v_0\rangle_1 = 2$.
(ii) $Y = e_3 + \sqrt{2}\, e_5$ lies in the totally umbilic hypersurface

$$N_{v_0,0} = \{p \in R^{4,1} \mid \langle p,p\rangle_1 = -1, \langle p, v_0\rangle_1 = 0\}$$

of H^4 and $N_{v_0,0}$ is isometric to H^3, where H^n is the hyperbolic space form $N^n(-1)$.
(iii) Y has constant mean curvature 2 in H^3.

PROOF. Let ω_{ij} denote the (i,j)-th entry of θ^{II}_λ at $\lambda = \sqrt{2}$. Since $dE = E\theta^{II}_\lambda$, we have

$$de_j = \sum_{i=1}^{5} \omega_{ij} e_i, \quad \omega_{ij} + \epsilon_i\epsilon_j\omega_{ji} = 0 \tag{8.10}$$

for $1 \le j \le 5$, where $\epsilon_1 = \cdots = \epsilon_4 = -\epsilon_5 = 1$. Read ω_{ij} from (4.5) to get

$$\begin{aligned} &\omega_{12} = u_y\, dx - u_x\, dy, \\ &\omega_{13} = e^{-u} dx, \quad \omega_{14} = -\sqrt{2}\cosh u\, dx, \quad \omega_{15} = \sqrt{2}\sinh u\, dx, \\ &\omega_{23} = -e^{-u}\, dy, \quad \omega_{24} = -\sqrt{2}\sinh u\, dy, \quad \omega_{25} = \sqrt{2}\cosh u\, dy, \\ &\omega_{ij} = 0, \quad \text{if } 3 \le i,j \le 5. \end{aligned} \tag{8.11}$$

To prove (i), we compute directly

$$\begin{aligned} dv_0 &= \sqrt{2}\, de_3 + de_4 + de_5, \\ &= \sqrt{2}\,\omega_{13}e_1 + \sqrt{2}\,\omega_{23}e_2 + \omega_{14}e_1 + \omega_{24}e_2 + \omega_{15}e_1 + \omega_{25}e_2 \\ &= (\sqrt{2}\,\omega_{13} + \omega_{14} + \omega_{15})e_1 + (\sqrt{2}\,\omega_{23} + \omega_{24} + \omega_{25})e_2. \end{aligned}$$

Substitute (8.11) to the above equation to conclude that $dv_0 = 0$.

By Proposition 7.1, $N_{v_0,0}$ has constant sectional curvature -1. A direct computation gives (ii).

Use (8.10), (8.11) and a direct computation to get

$$dY = \frac{e^u}{\sqrt{2}}\,(dx\ e_1 + dy\ e_2).$$

So e_1, e_2 are tangent to Y and the dual 1-frame is

$$\omega_1 = \frac{1}{\sqrt{2}}\,e^u dx, \quad \omega_2 = \frac{1}{\sqrt{2}}\,e^u dy.$$

The normal plane of $N_{v_0,0}$ in $R^{4,1}$ at Y is spanned by Y and $v_0 = \sqrt{2}\,e_3 + e_4 + e_5$. It is easy to write down an $O(4,1)$-frame $(e_1, e_2, \tilde{e}_3, \tilde{e}_4, \tilde{e}_5)$ on Y such that $\tilde{e}_3$ is normal to Y in $N_{v_0,0}$ and $\tilde{e}_5 = Y$:

$$\begin{aligned}
\tilde{e}_3 &= e_3 - \frac{1}{\sqrt{2}}\,e_4 + \frac{1}{\sqrt{2}}\,e_5,\\
\tilde{e}_4 &= (e_3 + \frac{1}{\sqrt{2}}\,e_4 + \frac{1}{\sqrt{2}}\,e_5),\\
\tilde{e}_5 &= Y = e_3 + \sqrt{2}\,e_5.
\end{aligned}$$

Let $\tilde{e}_i = e_i$ for $1 \le i \le 2$, and $\tilde{\omega}_{ij} = \langle d\tilde{e}_i, \tilde{e}_j\rangle_1$. A direct computation gives

$$\tilde{\omega}_{13} = (e^{-u} + e^u)dx, \quad \tilde{\omega}_{23} = (-e^{-u} + e^u)dy.$$

So Y has constant mean curvature 2 in H^3. ∎

9. Loop group action for $G_{m,n}$-systems

In this section, we give an explicit construction of the dressing action of a rational map with two simple poles on the space of solutions of the $G_{m,n}$-systems.

First we review the construction of the dressing action. Let U/K be a symmetric space, and

$$\begin{aligned}
G_+ = \{g : \boldsymbol{C} \to U_{\boldsymbol{C}} \mid\ & g \text{ is holomorphic, satisfies the}\\
& U/K\text{-reality condition (3.5)}\},\\
G_- = \{g : S^2 \to U_{\boldsymbol{C}} \mid\ & g(\infty) = I,\ g \text{ is meromorphic, and}\\
& \text{satisfies the } U/K\text{-reality condition (3.5)}\}.
\end{aligned}$$

Let $g \in G_-$, v a solution of the U/K-system (1.1), and $E(x,\lambda)$ the frame of v with $E(0,\lambda) = I$, i.e.,

$$E^{-1}dE = \theta_\lambda, \quad E(0,\lambda) = I,$$

where θ_λ is defined by (2.3). Since θ_λ is holomorphic for $\lambda \in \mathbf{C}$, $E(x,\cdot) \in G_+$. It is known that (cf. [TU2]) $g(\lambda)E(x,\lambda)$ can be factored as

$$g(\lambda)E(x,\lambda) = \tilde{E}(x,\lambda)\tilde{g}(x,\lambda) \tag{9.1}$$

such that $\tilde{E}(x,\cdot) \in G_+$ and $\tilde{g}(x,\cdot) \in G_-$ for x in a neighborhood of the origin. Moreover, this factorization (9.1) can be obtained explicitly using residue calculus, and we have:

(i) $\tilde{E}$ is the frame for a new solution $\tilde{v}$ of the U/K-system with $\tilde{E}(0,\lambda) = I$.
(ii) Write $\tilde{g}(x,\lambda) = I + \lambda^{-1}m_1(x) + \lambda^{-2}m_2(x) + \cdots$. Then $\tilde{v} = v - p_1(m_1)$ is a new solution of the U/K-system, where p_1 is the projection onto $\mathcal{P} \cap \mathcal{A}^\perp$.
(iii) $g \sharp v = \tilde{v}$ defines an action of G_- on the space of germs of solutions of the U/K-system.
(iv) If U is compact and v is a smooth solution decaying at infinity, then the factorization (9.1) can be carried out for all $x \in R^n$ and the solution $g\sharp v$ is globally defined.
(v) Let $\tilde{\theta}$ denote the Lax connection of the solution $\tilde{v}$. Then

$$\tilde{g}\theta - d\tilde{g} = \tilde{\theta}\tilde{g}, \tag{9.2}$$

and this gives a system of compatible ordinary differential equations for $\tilde{v}$.

9.1 Remark. Let E be a frame (cf. Definition 3.3) of the solution v of the U/K-system (1.1). Then
(i) $g(\lambda)E(x,\lambda)$ is a frame of v if and only if g satisfies the U/K-reality condition,
(ii) if $g_0 \in U_{\mathbf{C}}$ is a constant, then $g_0E(x,\lambda)$ is a frame of v if and only if $g_0 \in K$.

Next we find certain simple element in G_- explicitly. The $G_{m,n}$- reality condition is

$$\begin{cases} \overline{g(\bar{\lambda})} = g(\lambda), \\ I_{m,n}g(\lambda)I_{m,n}^{-1} = g(-\lambda), \\ g(\lambda)g(\lambda)^t = I. \end{cases} \tag{9.3}$$

9.2 Remark. If $g(\lambda)$ satisfies the $G_{m,n}$- reality condition (9.3), then
(i) $g(0) \in O(m) \times O(n)$ and $g(\lambda) \in O(m+n, C)$ for all $\lambda \in C$,
(ii) $g(\lambda)$ satisfies the $U(n)$-reality condition, i.e.,

$$g(\bar{\lambda})^* g(\lambda) = I.$$

It is known (cf. [U]) that the group of rational maps $g : S^2 \to GL(n, C)$ satisfying the $U(n)$-reality condition $\overline{g(\bar{\lambda})}^t g(\lambda) = I$ is generated by the set of

$$h_{z,\pi}(\lambda) = \pi + \frac{\lambda - z}{\lambda - \bar{z}}(I - \pi), \tag{9.4}$$

where $z \in C$ and π is a Hermitian projection of C^n. Although $h_{z,\pi}$ defined by (9.4) does not satisfy the $G_{m,n}$- reality condition (9.3), a product of suitable choices of two such elements satisfies (9.3). To construct this element, let $R^m = \mathcal{M}_{m\times 1}$, W

and Z unit vectors in $\boldsymbol{R}^m$ and $\boldsymbol{R}^n$ respectively, and π the Hermitian projection of C^{n+m} onto $\boldsymbol{C}\begin{pmatrix} W \\ iZ \end{pmatrix}$, the complex linear subspace spanned by $\begin{pmatrix} W \\ iZ \end{pmatrix}$. So

$$\pi = \frac{1}{2}\begin{pmatrix} WW^t & -iWZ^t \\ iZW^t & ZZ^t \end{pmatrix}. \tag{9.5}$$

Note that $\bar{\pi}$ is the Hermitian projection onto $\boldsymbol{C}\begin{pmatrix} -W \\ iZ \end{pmatrix}$, which is perpendicular to $\begin{pmatrix} W \\ iZ \end{pmatrix}$. This implies

$$\pi\bar{\pi} = \bar{\pi}\pi = 0.$$

Let $s \in \boldsymbol{R}, s \neq 0$. Define

$$g_{s,\pi}(\lambda) = \left(\pi + \frac{\lambda - is}{\lambda + is}(I - \pi)\right)\left(\bar{\pi} + \frac{\lambda + is}{\lambda - is}(I - \bar{\pi})\right). \tag{9.6}$$

Substitute (9.5) to (9.6) to get

$$g_{s,\pi}(\lambda) = I + \frac{2s}{\lambda^2 + s^2}\begin{pmatrix} -sWW^t & \lambda WZ^t \\ -\lambda ZW^t & -sZZ^t \end{pmatrix}. \tag{9.7}$$

A direct computation implies that $g_{s,\pi}$ satisfies the $G_{m,n}$-reality condition (9.3). So $g_{s,\pi} \in G_-$. Note also that

$$g_{s,\pi}(0) = \begin{pmatrix} I - 2WW^t & 0 \\ 0 & I - 2ZZ^t \end{pmatrix}.$$

Below we give an explicit construction of the dressing action of $g_{s,\pi}$ on the space of solutions of the $G_{m,n}$-system (3.1).

9.3 Theorem. *Let $\xi : R^n \to \mathcal{M}_{m\times n}$ be a solution of the $G_{m,n}$-system (3.1), and $E(x,\lambda)$ a frame of ξ such that $E(x,\lambda)$ is holomorphic for $\lambda \in C$. Let W and Z be unit vectors in R^m, R^n respectively, π the Hermitian projection of $\boldsymbol{C}^{n+m}$ onto $\boldsymbol{C}\begin{pmatrix} W \\ iZ \end{pmatrix}$, and $g_{s,\pi}$ the map defined by (9.6). Let $\tilde{\pi}(x)$ denote the Hermitian projection onto $\boldsymbol{C}\begin{pmatrix} \tilde{W} \\ i\tilde{Z} \end{pmatrix}(x)$, where*

$$\begin{pmatrix} \tilde{W} \\ i\tilde{Z} \end{pmatrix}(x) = E(x, -is)^{-1}\begin{pmatrix} W \\ iZ \end{pmatrix}. \tag{9.8}$$

Let $\hat{W} = \tilde{W}/\parallel \tilde{W} \parallel$ and $\hat{Z} = \tilde{Z}/\parallel \tilde{Z} \parallel$,

$$\tilde{E}(x,\lambda) = g_{s,\pi}(\lambda)E(x,\lambda)g_{s,\tilde{\pi}(x)}(\lambda)^{-1}, \tag{9.9}$$

$$\tilde{\xi} = \xi - 2s(\hat{W}\hat{Z}^t)_*, \tag{9.10}$$

where $(\eta_*)_{ij} = \eta_{ij}$ if $i \neq j$, and $(\eta_*)_{ii} = 0$ if $1 \leq i \leq n$. Then $\tilde{\xi}$ is a solution of (3.1), $\tilde{E}$ is a frame for $\tilde{\xi}$ and $\tilde{E}(x,\lambda)$ is holomorphic in $\lambda \in C$.

When $m = n$, Theorem 9.3 was proved in [Zh].

To prove Theorem 9.3, we need the following lemma.

9.4 Lemma. *With the same assumption as in Theorem 9.3, the following statements are true:*

(i) $\tilde{W}(x) \in R^m$ *and* $\tilde{Z}(x) \in R^n$.

(ii) $\| \tilde{W}(x) \| = \| \tilde{Z}(x) \|$ *for all* x, *and* $g_{s,\tilde{\pi}(x)}$ *satisfies the* $G_{m,n}$*-reality condition (9.3), i.e., it belongs to* G_-.

(iii) $\tilde{E}(x, \lambda)$ *is holomorphic in* $\lambda \in \boldsymbol{C}$.

PROOF.

(i) Let A^* denote $\bar{A}^t$. Since $E(x, \lambda)$ satisfies the $G_{m,n}$- reality condition (9.3),

$$I_{m,n} E(x, -is)^{-1} I_{m,n}^{-1} = E(x, is)^{-1} = E(x, -is)^*. \tag{9.11}$$

Write $E(x, -is)^{-1} = \begin{pmatrix} \eta_1 & \eta_2 \\ \eta_3 & \eta_4 \end{pmatrix}$ with $\eta_1 \in gl(m, C)$ and $\eta_4 \in gl(n, C)$. It follows from (9.11) that η_1, η_4 are real and η_2, η_3 are pure imaginary. This implies that $\tilde{W}, \tilde{Z}$ are real.

(ii) Let $\langle v_1, v_2 \rangle = v_1^t v_2$ denote the standard bi-linear form on $\boldsymbol{C}^{n+m}$. Then $\langle u, u \rangle = 0$, where $u = \begin{pmatrix} W \\ iZ \end{pmatrix}$. Since E satisfies the reality condition (9.3), $E(x, \lambda) \in O(m + n, C)$. So

$$\langle E(x, -is)^{-1}(u), E(x, -is)^{-1}(u) \rangle = \langle u, u \rangle = 0.$$

This implies that $\| \tilde{W}(x) \| = \| \tilde{Z}(x) \|$ and $g_{s,\tilde{\pi}(x)}$ satisfies the reality condition (9.3).

(iii) If g satisfies (9.3), then $g(\lambda)^{-1} = g(\bar{\lambda})^*$. So

$$\begin{aligned} &\tilde{E}(x, \lambda) = g_{s,\pi}(\lambda) E(x, \lambda) g_{x,\tilde{\pi}(x)}(\bar{\lambda})^* = \\ &(\pi + \frac{\lambda - is}{\lambda + is} \pi^\perp)(\bar{\pi} + \frac{\lambda + is}{\lambda - is} \bar{\pi}^\perp) E(x, \lambda)(\bar{\tilde{\pi}} + \frac{\lambda - is}{\lambda + is} \bar{\tilde{\pi}}^\perp)(\tilde{\pi} + \frac{\lambda + is}{\lambda - is} \tilde{\pi}^\perp), \end{aligned} \tag{9.12}$$

where $\pi^\perp = I - \pi$ and $\tilde{\pi}^\perp = I - \tilde{\pi}$. The right hand side only has poles at $is, -is$ of order 2. First note that the coefficient of $\frac{1}{(\lambda - is)^2}$ on the right hand side of (9.12) is $-4s^2 \pi \bar{\pi}^\perp E(x, is) \bar{\tilde{\pi}} \tilde{\pi}^\perp$, which is equal to $-4s^2 \pi E(x, is) \bar{\tilde{\pi}}$ because

$$\pi \bar{\pi} = \bar{\pi} \pi = 0, \quad \tilde{\pi} \bar{\tilde{\pi}} = \bar{\tilde{\pi}} \pi = 0. \tag{9.13}$$

Claim that

$$\pi E(x, is) \bar{\tilde{\pi}} = 0. \tag{9.14}$$

To see this, we note that the image of $\pi E(x, is) \bar{\tilde{\pi}}$ is spanned by

$$\begin{aligned} \pi E(x, is) \begin{pmatrix} \tilde{W} \\ -i\tilde{Z} \end{pmatrix} &= \pi E(x, is) I_{m,n} \begin{pmatrix} \tilde{W} \\ i\tilde{Z} \end{pmatrix}, \quad \text{by (9.11)} \\ = \pi I_{m,n} E(x, -is) \begin{pmatrix} \tilde{W} \\ i\tilde{Z} \end{pmatrix} &= \pi I_{m,n} \begin{pmatrix} W \\ iZ \end{pmatrix} = \pi \begin{pmatrix} W \\ -iZ \end{pmatrix} = 0. \end{aligned}$$

This proves the claim. So the coefficient of $\frac{1}{(\lambda-is)^2}$ on the right hand side of (9.12) is zero.

The coefficient of $\frac{1}{\lambda-is}$ on the right hand side of (9.12) is $4is\pi E(x,is)\bar{\pi}$, which is zero because of (9.14). So $\tilde{E}(x,\lambda)$ is holomorphic at $\lambda = is$. Similar computation implies that $\tilde{E}(x,\lambda)$ is holomorphic at $\lambda = -is$. Hence $E(x,\lambda)$ is holomorphic for $\lambda \in \boldsymbol{C}$. ∎

9.5 Proof of Theorem 9.3.

A direct computation gives

$$\tilde{E}^{-1}d\tilde{E} = \tilde{g}E^{-1}dE\tilde{g}^{-1} - d\tilde{g}\tilde{g}^{-1}, \tag{9.15}$$

where $\tilde{g}(x,\lambda) = g_{s,\tilde{\pi}(x)}(\lambda)$. But $E^{-1}dE = \sum(a_i\lambda + [a_i, v])dx_i$, $\tilde{E}(x,\lambda)$ is holomorphic in $\lambda \in \boldsymbol{C}$ and $\tilde{g}(x,\lambda)$ is holomorphic at $\lambda = \infty$. So $\tilde{E}^{-1}d\tilde{E}$ must be of the form $\sum(a_i\lambda + \eta_i)dx_i$. Write

$$\tilde{g}(x,\lambda) = I + \lambda^{-1}m_1(x) + \cdots.$$

A direct computation shows that

$$m_1(x) = 2s\begin{pmatrix} 0 & \hat{W}\hat{Z}^t \\ -\hat{Z}\hat{W}^t & 0 \end{pmatrix}, \quad m_1(x) \in \mathcal{P}. \tag{9.16}$$

Multiply (9.15) by $\tilde{g}$ on the right to get

$$\left(\sum_i (a_i\lambda + \eta_i)dx_i\right)(1 + m_1\lambda^{-1} + \cdots)$$
$$= (I + m_1\lambda^{-1} + \cdots)\left(\sum_i (a_i\lambda + [a_i, v])dx_i\right) - (dm_1\lambda^{-1} + \cdots).$$

Equate the constant term of the above equation to get

$$\eta_i = [a_i, v - m_1] = [a_i, v - p_1(m_1)],$$

where p_1 is the projection from $\mathcal{P}$ onto $\mathcal{P} \cap \mathcal{A}^\perp$. Write $\tilde{v} = \begin{pmatrix} 0 & \tilde{\xi} \\ -\tilde{\xi}^t & 0 \end{pmatrix}$. Theorem follows from (9.16). ∎

Since $E(x,\lambda)$ in Theorem 9.3 is not assumed to satisfy the initial condition $E(0,\lambda) = I$, the resulting new solution $\tilde{\xi}$ is not necessarily equal to the dressing action of $g_{s,\pi}$ on ξ. But they are related as follows:

9.6 Corollary. *Suppose E is a frame of the solution ξ of the $G_{m,n}$-system (3.1) such that $E(x,\lambda)$ is holomorphic for $\lambda \in \boldsymbol{C}$.*

(i) If $E(0,\lambda) = I$, then $\tilde{\xi}$ obtained in Theorem 9.3 is $g_{s,\pi}\sharp\xi$ and $\tilde{E}$ is the frame of $\tilde{\xi}$ with $\tilde{E}(0,\lambda) = I$.

(ii) Let $g_+(\lambda) = E(0,\lambda)$, and $\tilde{\xi}$ the new solution of (3.1) obtained in Theorem 9.3. Then $g_+ \in G_+$ and $\tilde{\xi} = \tilde{g}_-\sharp\xi$, where $\tilde{g}_-$ is obtained by factoring $g_{s,\pi}g_+ = \tilde{g}_+\tilde{g}_-$ with $\tilde{g}_\pm \in G_\pm$.

The functions $\tilde{W}, \tilde{Z}$ in Theorem 9.3 also satisfy a system of compatible first order differential equations. This follows from taking the differential of the defining equation (9.8) of $\tilde{W}, \tilde{Z}$:

$$d\begin{pmatrix}\tilde{W}\\ i\tilde{Z}\end{pmatrix} = dE_{-is}^{-1}\begin{pmatrix}W\\ iZ\end{pmatrix} = -E_{-is}^{-1}dE_{-is}\begin{pmatrix}\tilde{W}\\ i\tilde{Z}\end{pmatrix} = -\theta_{-is}\begin{pmatrix}\tilde{W}\\ i\tilde{Z}\end{pmatrix}, \tag{9.17}$$

where θ_λ is defined by (3.2) and $E_\lambda(x) = E(x,\lambda)$. We write system (9.17) explicitly:

$$\begin{cases}\tilde{W}_{x_i} = (\xi D_i^t - D_i\xi^t)\tilde{W} + sD_i\tilde{Z},\\ \tilde{Z}_{x_i} = sD_i^t\tilde{W} + (\xi^t D_i - D_i^t\xi)\tilde{Z},\end{cases} \tag{9.18}$$

or

$$\begin{cases}(\tilde{w}_i)_{x_i} = -\sum_{j\neq i}^n f_{ji}\tilde{w}_j - \sum_{j=1}^{m-n} g_{ji}\tilde{w}_{j+n} + s\tilde{z}_i & i\leq n,\\ (\tilde{w}_i)_{x_j} = f_{ij}\tilde{w}_j & i\leq n, j\neq i,\\ (\tilde{w}_i)_{x_j} = g_{ij}\tilde{w}_j & i>n\ ,\\ (\tilde{z}_i)_{x_j} = f_{ji}\tilde{z}_j & j\neq i\ ,\\ (\tilde{z}_i)_{x_i} = -\sum_{j\neq i}^n f_{ij}\tilde{z}_j + s\tilde{w}_i. \end{cases} \tag{9.19}$$

The converse is also true:

9.7 Proposition. *Given $\xi : R^n \to \mathcal{M}^0_{m\times n}$ and a real number $s \neq 0$, we have:*

(i) System (9.18) is solvable for $\tilde{W}, \tilde{Z}$ if and only if ξ is a solution of the $G_{m,n}$-system (3.1).

(ii) Let ξ be a solution of (3.1), and $\tilde{W}, \tilde{Z}$ solution of (9.18) with initial condition $\tilde{W}(0) = W, \tilde{Z}(0) = Z$ for some unit vectors $W \in R^m, Z \in R^n$. Set

$$\tilde{\xi} = \xi - 2s\frac{(\tilde{W}\tilde{Z}^t)_*}{\|\tilde{W}\|\ \|\tilde{Z}\|},$$

where $(Y_*)_{ij} = Y_{ij}$ if $i \neq j$, and $(Y_*)_{ii} = 0$ if $1 \leq i \leq n$. Then $\tilde{\xi}$ is a solution of (3.1), and is equal to $g_{s,\pi}\sharp\xi$, where π is the projecton onto $C(W, iZ)^t$.

PROOF. System (9.18) is the same as system (9.17), i.e.,

$$d\begin{pmatrix}\tilde{W}\\ i\tilde{Z}\end{pmatrix} = -\theta_{-is}\begin{pmatrix}\tilde{W}\\ i\tilde{Z}\end{pmatrix}, \tag{9.20}$$

where

$$\theta_{-is} = \sum_{j=1}^n \begin{pmatrix} D_j\xi^t - \xi D_j^t & isD_j\\ -isD_j & -\xi^t D_j + D_j^t\xi\end{pmatrix} dx_j.$$

But (9.20) is solvable if and only if θ_{-is} is flat. This is equivalent to ξ being a solution of (3.1). So there exists a unique solution $\tilde{W}, \tilde{Z}$ of (9.18) such that $\tilde{W}(0) = W$ and $\tilde{Z}(0) = Z$. Let $E(x,\lambda)$ be the frame of ξ with $E(0,\lambda) = I$. Then

$E_{-is}^{-1}\begin{pmatrix} W \\ iZ \end{pmatrix}$ is also a solution of (9.18) with the same initial condition. Hence they must be equal and Proposition follows from Theorem 9.3. ∎

Since the Lax connection of the $G_{m,n}$-system I, II are gauge equivalent to the Lax connection of the $G_{m,n}$-system, we can write down the action of $g_{s,\pi}$ on solutions of these two systems easily from the action on solutions of the $G_{m,n}$-system.

Use the same notations as in Theorem 9.3. Let

$$\begin{aligned} \tilde{E}^{\natural}(x,\lambda) &= g_{s,\pi}^{-1}\tilde{E}(x,\lambda) = E(x,\lambda)g_{s,\tilde{\pi}(x)}^{-1} \\ &= E(x,\lambda)\begin{pmatrix} I_m - \frac{2s^2}{\lambda^2+s^2}\hat{W}\hat{W}^t & -\frac{2s\lambda}{\lambda^2+s^2}\hat{W}\hat{Z}^t \\ \frac{2s\lambda}{\lambda^2+s^2}\hat{Z}\hat{W}^t & I_n - \frac{2s^2}{\lambda^2+s^2}\hat{Z}\hat{Z}^t \end{pmatrix}. \end{aligned} \tag{9.21}$$

Since both $g_{s,\pi}$ and $\tilde{E}$ satisfy the $G_{m,n}$- reality condition (9.3), so does $\tilde{E}^{\natural}$. Hence $\tilde{E}^{\natural}$ is a frame of $\tilde{\xi}$. Note that $\tilde{E}^{\natural}(x,\cdot)$ is not in G_+. The $G_{m,n}$- reality condition implies that both $E(x,0)$ and $\tilde{E}^{\natural}(x,0)$ are in $O(m)\times O(n)$. Write

$$E(x,0) = \begin{pmatrix} A(x) & 0 \\ 0 & B(x) \end{pmatrix}, \quad \tilde{E}^{\natural}(x,0) = \begin{pmatrix} \tilde{A}^{\natural}(x) & 0 \\ 0 & \tilde{B}^{\natural}(x) \end{pmatrix}.$$

It follows from (9.21) that we have

$$\begin{cases} \tilde{A}^{\natural} = A(I - 2\hat{W}\hat{W}^t), \\ \tilde{B}^{\natural} = B(I - 2\hat{Z}\hat{Z}^t). \end{cases} \tag{9.22}$$

Write

$$\xi = \begin{pmatrix} F \\ G \end{pmatrix}, \quad \tilde{\xi} = \begin{pmatrix} \tilde{F} \\ \tilde{G} \end{pmatrix}, \quad A = (A_1, A_2), \quad \tilde{A}^{\natural} = (\tilde{A}_1^{\natural}, \tilde{A}_2^{\natural}),$$

where $A_1, \tilde{A}_1^{\natural} \in \mathcal{M}_{m\times n}$ and $A_2, \tilde{A}_2^{\natural} \in \mathcal{M}_{m\times(m-n)}$. Rewrite (9.10) as

$$\begin{pmatrix} \tilde{F} \\ \tilde{G} \end{pmatrix} = \begin{pmatrix} F \\ G \end{pmatrix} - 2s\left(\hat{W}\hat{Z}^t\right)_*.$$

So (A_1, F) and $(\tilde{A}_1^{\natural}, \tilde{F})$ are solutions of the $G_{m,n}$-system I (3.4), and (F, G, B) and $(\tilde{F}, \tilde{G}, \tilde{B}^{\natural})$ are solutions of the $G_{m,n}$-system II (3.10). Recall that

(i) θ_λ^I defined by (3.3) is the Lax connection of the solution (A_1, F) of the $G_{m,n}$-system I (3.4),

(ii) θ_λ^{II} defined by (3.9) is the Lax connection of solution (F, G, B) of the $G_{m,n}$-system II (3.10),

(iii) θ_λ^I is the gauge transformation of θ_λ by $\begin{pmatrix} A & 0 \\ 0 & I_n \end{pmatrix}$, and θ_λ^{II} is the gauge transformation of θ_λ by $\begin{pmatrix} I_m & 0 \\ 0 & B \end{pmatrix}$.

So the frames E of ξ, $\tilde{E}^\natural$ of $\tilde{\xi}$, E^I of (A_1, F), $\tilde{E}^{\natural I}$ of $(\tilde{A}_1^\natural, \tilde{F})$ and E^{II} of (F, G, B) and $\tilde{E}^{\natural II}$ of $(\tilde{F}, \tilde{G}, \tilde{B}^\natural)$ are related by

$$\begin{aligned} E^I(x,\lambda) &= E(x,\lambda)\begin{pmatrix} A^t & 0 \\ 0 & I_n \end{pmatrix}, \\ \tilde{E}^{\natural I}(x,\lambda) &= \tilde{E}^\natural(x,\lambda)\begin{pmatrix} \tilde{A}^{\natural t} & 0 \\ 0 & I_n \end{pmatrix}, \\ \tilde{E}^{\natural II}(x,\lambda) &= \tilde{E}^\natural(x,\lambda)\begin{pmatrix} I_m & 0 \\ 0 & \tilde{B}^{\natural t} \end{pmatrix}. \end{aligned}$$

Use (9.21) to get

$$\begin{aligned} \tilde{E}^{\natural I}(x,\lambda) &= E^I(x,\lambda)\left(I - \frac{2}{\lambda^2+s^2}\begin{pmatrix} \lambda^2 A\hat{W}\hat{W}^t A^t & s\lambda A\hat{W}\hat{Z}^t \\ s\lambda\hat{Z}\hat{W}^t A^t & s^2\hat{Z}\hat{Z}^t \end{pmatrix}\right), \\ &= E^I(x,\lambda)\left(I - \frac{2}{\lambda^2+s^2}\begin{pmatrix} \lambda A\hat{W} \\ s\hat{Z} \end{pmatrix}\begin{pmatrix} \lambda\hat{W}^t A^t & s\hat{Z}^t \end{pmatrix}\right), \end{aligned} \tag{9.23}$$

and

$$\tilde{E}^{\natural II}(x,\lambda) = E^{II}(x,\lambda)\left(I - \frac{2}{\lambda^2+s^2}\begin{pmatrix} s^2\hat{W}\hat{W}^t & -s\lambda\hat{W}\hat{Z}^t B^t \\ -s\lambda B\hat{Z}\hat{W}^t & \lambda^2 B\hat{Z}\hat{Z}^t B^t \end{pmatrix}\right). \tag{9.24}$$

It follows from Corollary 9.6 that the new solution $\tilde{\xi}$ obtained in Theorem 9.3 depends on $g_{s,\pi}$ and the frame E. Henceforth, we will use the following notations:

$$\begin{aligned} &(\tilde{\xi}, \tilde{E}^\natural) = g_{s,\pi}\cdot(\xi, E), \\ &\tilde{A}^\natural = g_{s,\pi}\cdot A, \quad \tilde{B}^\natural = g_{s,\pi}\cdot B, \\ &(\tilde{A}_1^\natural, \tilde{F}, \tilde{E}^{\natural I}) = g_{s,\pi}\cdot(A_1, F, E^I), \\ &(\tilde{F}, \tilde{G}, \tilde{B}^\natural, \tilde{E}^{\natural II}) = g_{s,\pi}\cdot(F, G, B, E^{II}). \end{aligned}$$

Use exactly the same argument as for Theorem 9.3 to get the action of $g_{s,\pi}$ on solutions of the partial $G_{m,n+1}$-system (3.12). We summarize the results for this case below.

9.8 Theorem. *Let (F, G, b) be a solution of the partial $G_{m,n+1}$-system (3.12), Θ_λ (defined by (3.11)) its Lax connection, and $E(x,\lambda)$ a frame of (F, G, b). Let $W \in R^m$ and $Z \in R^{n+1}$ be unit vectors, π the Hermitian projection onto $\mathbf{C}\begin{pmatrix} W \\ iZ \end{pmatrix}$, and $g_{s,\pi}$ defined by (9.6). Let*

$$\begin{pmatrix} \tilde{W} \\ i\tilde{Z} \end{pmatrix}(x) = E(x, -is)^{-1}\begin{pmatrix} W \\ iZ \end{pmatrix}, \tag{9.25}$$

$\tilde{\pi}(x)$ the Hermitian projection onto $\boldsymbol{C}\begin{pmatrix}\tilde{W}\\ i\tilde{Z}\end{pmatrix}(x)$,

$$\tilde{E}(x,\lambda) = g_{s,\pi}(\lambda)E(x,\lambda)g^{-1}_{s,\tilde{\pi}(x)}(\lambda),$$
$$\begin{pmatrix}\tilde{F} & \tilde{b}\\ \tilde{G} & 0\end{pmatrix} = \begin{pmatrix}F & b\\ G & 0\end{pmatrix} - 2s(\hat{W}\hat{Z}^t)_*, \tag{9.26}$$

where $\hat{W} = \tilde{W}/\parallel \tilde{W}\parallel$, $\hat{Z} = \tilde{Z}/\parallel \tilde{Z}\parallel$. Then:

(i) $(\tilde{F},\tilde{G},\tilde{b})$ is a solution of the partial $G_{m,n+1}$-system (3.12) and $\tilde{E}$ is a frame.

(ii) $(\tilde{W},\tilde{Z})$ is a solution of

$$\begin{cases}(\tilde{W})_{x_j} = -\begin{pmatrix}-FC_j + C_jF^t & C_jG^t\\ -GC_j & 0\end{pmatrix}\tilde{W} + sC_j\tilde{Z},\\ (\tilde{Z})_{x_j} = -\begin{pmatrix}-F^tC_j + C_jF & C_jb\\ -b^tC_j & 0\end{pmatrix}\tilde{Z} + s(C_j,0)\tilde{W}.\end{cases} \tag{9.27}$$

(iii) System (9.27) is solvable for $(\tilde{W},\tilde{Z})$ if and only if (F,G,b) is a solution of (3.12).

(iv) If (F,G,b) is a solution of (3.12) and $(\tilde{W},\tilde{Z})$ is a solution of (9.27) such that $\tilde{W}(0),\tilde{Z}(0)$ are unit real vectors, then $(\tilde{F},\tilde{G},\tilde{b})$ defined by formula (9.26) is a solution of (3.12).

Let (F,G,b) be a solution of the partial $G_{m,n+1}$-system (3.12), and E a frame of (F,G,b). Since $E(x,0)\in O(m)\times O(n+1)$, there exist $A(x)\in O(m)$, and $B(x)\in O(n+1)$ such that

$$E(x,0) = \begin{pmatrix}A(x) & 0\\ 0 & B(x)\end{pmatrix}.$$

Write $A=(A_1,A_2)$ with $A_1\in\mathcal{M}_{m\times n}$ and $A_2\in\mathcal{M}_{m\times(m-n)}$. Then (A_1,F,b) is a solution of the partial $G_{m,n+1}$-system I (3.14) and

$$E^I(x,\lambda) = E(x,\lambda)\begin{pmatrix}A^t & 0\\ 0 & I_n\end{pmatrix}$$

is a frame of (A_1,F,b). Let $(\tilde{F},\tilde{G},\tilde{b})$, $\hat{W},\hat{Z}$, and $\tilde{E}$ be as in Theorem 9.8, and

$$\tilde{E}^\natural = g^{-1}_{s,\pi}\tilde{E}, \quad \tilde{A}^\natural = A(I-2\hat{W}\hat{W}^t).$$

Then $(\tilde{A}^\natural{}_1,\tilde{F},\tilde{b})$ is a new solution of the partial $G_{m,n+1}$-system I, where $\tilde{A}^\natural = (\tilde{A}^\natural_1,\tilde{A}^\natural_2)$. Moreover,

$$\tilde{E^\natural}^I(x,\lambda) = E^I(x,\lambda)\left(I - \frac{2}{\lambda^2+s^2}\begin{pmatrix}\lambda\, A\hat{W}\\ s\hat{Z}\end{pmatrix}\begin{pmatrix}\lambda\hat{W}^tA^t & s\hat{Z}^t\end{pmatrix}\right)$$

is a frame of $(\tilde{F},\tilde{G},\tilde{b})$. We will use the following notations:

$$(\tilde{F},\tilde{G},\tilde{b},\tilde{E}^\natural) = g_{s,\pi}\cdot(F,G,b,E),$$
$$(\tilde{A}^\natural_1,\tilde{F},\tilde{b},\tilde{E^\natural}^I) = g_{s,\pi}\cdot(A_1,F,b,E^I).$$

10. Ribaucour Transformations for $G_{m,n}$-systems

The main goal of this section is to give geometric interpretations of the action of $g_{s,\pi}$ on the spaces of solutions of the various $G_{m,n}$-systems constructed in section 9.

We first review some notions in classical differential geometry (cf. [Da], [Bi1], [Bi2]). Given a two parameter family of spheres $S(x,y)$ in R^3,

$$S(x,y): \quad p(x,y) + r(x,y)w, \quad w \in S^2,$$

generically there exist two enveloped surfaces $M, \tilde{M}$, i.e., they are tangent to these spheres. To see this, we fix a parametrization $w(u,v)$ on S^2. To find an enveloped surface is to find $u(x,y), v(x,y)$ so that $w(u(x,y), v(x,y))$ is normal to the surface

$$X(x,y) = p(x,y) + r(x,y)w(u(x,y), v(x,y))$$

at $X(x,y)$. So u, v need to satisfy $X_x \cdot w = 0$ and $X_y \cdot w = 0$. Or equivalently,

$$p_x \cdot w + r_x = 0, \quad p_y \cdot w + r_y = 0. \tag{10.1}$$

This means that $w(u(x,y), v(x,y))$ must lie in the intersection of the two circles defined by (10.1) on S^2. Since generically two such circles intersect at exactly two points for each (x,y), we obtain two surfaces $M, \tilde{M}$ and a map $\ell : M \to \tilde{M}$ so that $X(x,y)$ and $\tilde{X}(x,y)$ lie in the same sphere $S(x,y)$. Note that the map ℓ is characterized by the property that the normal line at q to M and the normal line at $\ell(q) = \tilde{q}$ to $\tilde{M}$ meet at equidistance $r(q)$. We call such map ℓ a *sphere congruence*. A sphere congruence $\ell : M \to \tilde{M}$ is called a *Ribaucour transformation* if it preserves line of curvature directions and the lines $p+te$ and $\ell(p)+t\ell_*(e)$ meet at equidistance for any principal direction $e \in TM_p$ and $p \in M$.

Natural generalizations of sphere congruence and Ribaucour transformation to submanifolds in space forms are given in [DT]. For $x \in N^m(c)$ and $v \in TN^m(c)_x$, let $\gamma_{x,v}(t) = \exp(tv)$ denote the geodesic.

10.1 Definition ([DT]). Let M^n and $\tilde{M}^n$ be submanifolds of the space form $N^m(c)$. A *sphere congruence* is a vector bundle isomorphism $P : \nu(M) \to \nu(\tilde{M})$ that covers a diffeomorphism $\ell : M \to \tilde{M}$ with the following properties:

(a) If ξ is a parallel normal vector field of M, then $P \circ \xi \circ \ell^{-1}$ is a parallel normal field of $\tilde{M}$.

(b) For any nonzero vector $\xi \in \nu_x(M)$, the geodesics $\gamma_{x,\xi}$ and $\gamma_{\ell(x),P(\xi)}$ intersect at a point that is equidistant from x and $\ell(x)$ (the distance depends on x).

10.2 Definition ([DT]). A sphere congruence $P : \nu(M) \to \nu(\tilde{M})$ that covers $\ell : M \to \tilde{M}$ is called a *Ribaucour Transformation* if it satisfies the following additional properties:

(i) If e is an eigenvector of the shape operator A_ξ of M, then $\ell_*(e)$ is an eigenvector of the shape operator $A_{P(\xi)}$ of $\tilde{M}$.

(ii) the geodesics $\gamma_{x,e}$ and $\gamma_{\ell(x),\ell_*(e)}$ intersect at a point equidistant to x and $\ell(x)$.

Below we show that the action of $g_{s,\pi}$ on solutions of the $G_{m,n}$-system I corresponds to a Ribaucour transformation for flat submanifolds in S^{n+m-1}.

10.3 Theorem. *Let E^I be a frame of the solution (A_1, F) of the $G_{m,n}$-system I (3.4), $g_{s,\pi}$ given by (9.6), and*

$$(\tilde{A}_1^\natural, \tilde{F}, \tilde{E^\natural}^I) = g_{s,\pi} \cdot (A_1, F, E^I)$$

as in section 9. Write

$$\begin{aligned} E^I(x,1) &= (X(x), e_{n+2}(x), \cdots, e_{n+m}(x), e_1(x), \cdots, e_n(x)), \\ \tilde{E^\natural}^I(x,1) &= (\tilde{X}(x), \tilde{e}_{n+2}(x), \cdots, \tilde{e}_{n+m}(x), \tilde{e}_1(x), \cdots, \tilde{e}_n(x)). \end{aligned} \tag{10.2}$$

Then:

(1) Both X and $\tilde{X}$ are immersions of flat n-dimensional submanifolds of S^{n+m-1} with flat, non-degenerate normal bundle, $x_1, \cdots, x_n$ line of curvature coordinates, $\{e_\alpha\}_{\alpha=n+2}^{n+m}$ and $\{\tilde{e}_\alpha\}_{\alpha=n+2}^{n+m}$ are parallel normal frames for X and $\tilde{X}$ respectively.

(2) (A_1, F) and $(\tilde{A}_1^\natural, \tilde{F})$ are solutions of (3.4) corresponding to X and $\tilde{X}$ as in Theorem 6.1 respectively.

(3) The bundle morphism $P : \nu(M) \to \nu(\tilde{M})$ defined by $P(e_\alpha(x)) = \tilde{e}_\alpha(x)$ for $n+2 \leq \alpha \leq n+m$ is a Ribaucour Transformation covering the map $X(x) \mapsto \tilde{X}(x)$.

PROOF.

(1) and (2) follow from Theorems 6.1 and 9.3.

(3) Let A_2, G and $A = (A_1, A_2)$ be given as in Proposition 3.6, and $\hat{W}, \hat{Z}$ as in Theorem 9.3. Let

$$\gamma = (\gamma_{n+1}, \ldots, \gamma_{m+n}, \gamma_1, \ldots, \gamma_n) = \begin{pmatrix} \sin\rho\, \hat{W}^t A^t & \cos\rho\, \hat{Z}^t \end{pmatrix},$$

where $\sin\rho = 1/\sqrt{1+s^2}$ and $\cos\rho = s/\sqrt{1+s^2}$. Substitute $\lambda = 1$ in (9.23) to get

$$\tilde{E^\natural}^I(x,1) = E^I(x,1)\left(I - 2\begin{pmatrix} \sin\rho\, A\hat{W} \\ \cos\rho\, \hat{Z} \end{pmatrix}\begin{pmatrix} \sin\rho\, \hat{W}^t A^t & \cos\rho\, \hat{Z}^t \end{pmatrix}\right). \tag{10.3}$$

Substitute (10.2) to the above equation to get formulas for each column of $\tilde{E^\sharp}^I$:

$$\tilde{X} = X(1 - 2\gamma_{n+1}^2) - 2\gamma_{n+1}\left(\sum_{j=2}^m \gamma_{n+j}e_{n+j} + \sum_{j=1}^n \gamma_j e_j\right),$$

$$\tilde{e}_i = -2\gamma_{n+1}\gamma_i X - 2\gamma_i\left(\sum_{j=2}^m \gamma_{n+j}e_{n+j} + \sum_{j=1}^m \gamma_j e_j\right) + e_i,$$

$$\tilde{e}_{n+i} = -2\gamma_{n+1}\gamma_{n+i} X - 2\gamma_{n+i}\left(\sum_{j=2}^m \gamma_{n+j}e_{n+j} + \sum_{j=1}^m \gamma_j e_j\right) + e_{n+i}.$$

So we have

$$\begin{aligned}\gamma_i\tilde{X} - \gamma_{n+1}\tilde{e}_i &= \gamma_i X - \gamma_{n+1}e_i,\\ \gamma_{n+i}\tilde{X} - \gamma_{n+1}\tilde{e}_i &= \gamma_{n+i}X - \gamma_{n+1}e_i.\end{aligned}$$

Let

$$\psi_i = \arctan(\gamma_{n+1}/\gamma_i), \quad \text{and} \quad \phi_i = \arctan(\gamma_{n+1}/\gamma_{n+i}).$$

Then the above equations are

$$\begin{aligned}\cos\psi_i\, X - \sin\psi_i\, e_i &= \cos\psi_i\,\tilde{X} - \sin\psi_i\,\tilde{e}_i,\\ \cos\phi_i X - \sin\phi_i e_{n+i} &= \cos\phi_i\tilde{X} - \sin\phi_i\tilde{e}_{n+i}.\end{aligned} \tag{10.4}$$

Geometrically, this means that the geodesic of S^{n+m-1} at $X(x)$ in the direction $e_i(x)$ intersects the geodesic of S^{n+m-1} at $\tilde{X}(x)$ in the direction $\tilde{e}_i(x)$ at a point equidistant to $X(x)$ and $\tilde{X}(x)$. So P is a Ribaucour Transformation. ∎

Note that the distance ϕ_i and ψ_i's in (10.4) in the proof of Theorem 10.3 satisfy

$$\frac{\sum_{i=1}^{n}\cot^2\psi_i(x)}{1+\sum_{i=2}^{m}\cot^2\phi_i(x)} = s^2,$$

for all x.

The following theorem gives the geometric transformation for flat submanifolds in R^{n+m} corresponding to the action of $g_{s,\pi}$ on the space of solutions of the $G_{m,n}$-system I. The proof is similar to that of Theorem 10.3.

10.4 Theorem. *Let X be a local isometric immersion of R^n in R^{n+m} with flat and non-degenerate normal bundle, and $(x_1, \cdots, x_n)$ a line of curvature coordinate system. Let (A_1, F) be the corresponding solution of the $G_{m,n}$-system I (3.4), b, g as in Theorem 6.3 (i) and (ii), and E^I a frame of (A_1, F) with $E^I(x,1) = g(x)$. Let A, G be as in Proposition 3.6, and $E = E^I\begin{pmatrix} A & 0\\ 0 & I_n\end{pmatrix}$ a frame for the solution $\xi = \begin{pmatrix} F\\ G\end{pmatrix}$ of (3.1). Let $g_{s,\pi}$, $\tilde{W}, \tilde{Z}$, $\hat{W} = \tilde{W}/\parallel\tilde{W}\parallel$, $\hat{Z} = \tilde{Z}/\parallel\tilde{Z}\parallel$ be as in Theorem 9.3 for the solution ξ, E a frame of ξ, and*

$$(\tilde{F}, \tilde{G}, \tilde{E^\natural}) = g_{s,\pi}\cdot(F, G, E), \quad (\tilde{A_1^\natural}^{I}, \tilde{F}, \tilde{E^\natural}^{I}) = g_{s,\pi}\cdot(A_1, F, E^I)$$

as in section 9. Write

$$\begin{aligned}E^I(x,1) &= (e_{n+1}(x), \cdots, e_{n+m}(x), e_1(x), \cdots, e_n(x)),\\ \tilde{E^\natural}^{I}(x,1) &= (\tilde{e}_{n+1}(x), \cdots, \tilde{e}_{n+m}(x), \tilde{e}_1(x), \cdots, \tilde{e}_n(x)).\end{aligned}$$

Set

$$\begin{aligned}\gamma = (\gamma_{n+1}, \ldots, \gamma_{m+n}, \gamma_1, \ldots, \gamma_n) &:= \left(\sin\rho\,\hat{W}^t A^t \quad \cos\rho\hat{Z}^t\right),\\ \eta(x) = E^I(x,1)\gamma(x)^t, &\quad \text{where } \rho = \cot^{-1}s.\end{aligned}$$

Then:

(1) There exists ϕ such that $\phi_{x_i} = -sb_i\tilde{z}_i$ for all $i = 1, \ldots, n$, where $\tilde{z}_i$ is the i-th coordinate of $\tilde{Z}$.

(2) Let $\tilde{X} = X + \frac{2\cos\rho\,\phi}{||\tilde{W}||}\eta$. Then $\tilde{X}$ is again an immersed flat submanifold in R^{n+m} with flat and non-degenerate normal bundle.

(3) The bundle morphism $P(e_\alpha(x)) = \tilde{e}_\alpha(x)$ for $n+1 \leq \alpha \leq n+m$ is a Ribaucour Transformation covering the map $X(x) \mapsto \tilde{X}(x)$.

(4) The solution of (3.4) corresponding to $\tilde{X}$ is $(\tilde{A}_1^\natural, \tilde{F})$.

PROOF.

(1) By (9.19), we have $(\tilde{z}_j)_{x_i} = f_{ij}\tilde{z}_j$. Theorem 6.3 implies that $(b_i)_{x_j} = f_{ij}b_j$ for $i \neq j$. A direct computation gives $(b_i\tilde{z}_i)_{x_j} = (b_j\tilde{z}_j)_{x_i}$. So ϕ exists.

(2) It follows from the definition of η and equation (9.19) for $\tilde{w}_i, \tilde{z}_j$ that

$$d\eta = -s\left(\sum_{i=1}^n \hat{z}_i\hat{w}_i dx_i\right)\eta + \frac{s}{\cos\rho}\sum_{i=1}^n \hat{w}_i dx_i e_i$$

and

$$d\left(\frac{1}{||\tilde{W}||}\right) = -\frac{s}{||\tilde{W}||}\left(\sum_{i=1}^n \hat{z}_i\hat{w}_i dx_i\right).$$

A straightforward calculation gives

$$d\tilde{X} = dX + \frac{2\cos\rho}{s}d\left(\frac{\phi}{||\tilde{W}||}\eta\right) = \sum_i \tilde{b}_i dx_i\tilde{e}_i,$$

where $\tilde{b}_i = b_i + \frac{2\phi\hat{w}_i}{||\tilde{W}||}$. Hence $\tilde{X}$ is a submanifold with the stated properties.

(3) Set $\lambda = 1$ in equation (9.23) to get

$$\tilde{e}_i = e_i - 2\gamma_i\eta.$$

So we have

$$\tilde{X} + \frac{\cos\rho\phi}{\gamma_i s||\tilde{W}||}\tilde{e}_i = X + \frac{\cos\rho\phi}{\gamma_i s||\tilde{W}||}e_i,$$

which implies that P is a Ribaucour Transformation covering $X \mapsto \tilde{X}$.

(4) follows from (2) and (3). ∎

The following theorem gives the geometric transformation for local isometric immersions of S^n in S^{m+n} corresponding to the action of $g_{s,\pi}$ on the space of solutions of the partial $G_{m,n+1}$-system I.

10.5 Theorem. *Let X be a local isometric immersion of S^n in S^{n+m} with flat and non-degenerate normal bundle, and $(x_1, \cdots, x_n)$ a line of curvature coordinate system. Let (A_1, F, b) be the solution of the partial $G_{m,n+1}$-system I (3.14) corresponding to X,*

$$g = (e_{n+1}, \cdots, e_{n+m}, e_1, \cdots, e_n, X)$$

as in Theorem 6.5, E^I the frame of (A_1, F, b) such that $E^I(x, 1) = g(x)$, and

$$(\tilde{A}_1^\natural, \tilde{F}, \tilde{b}, \tilde{E^\natural}^I) = g_{s,\pi} \cdot (A_1, F, b, E^I)$$

as in section 9. Write

$$\tilde{E^\natural}^I(x, \lambda) = (\tilde{e}_{n+1}(x), \cdots, \tilde{e}_{n+m}(x), \tilde{e}_1(x), \cdots, \tilde{e}_n(x), \tilde{X}(x)).$$

Then:

(i) $\tilde{X}$ is a local isometric immersion of S^n in S^{n+m} with flat and non-degenerate normal bundle, x is line of curvature coordinates, and $\{\tilde{e}_\alpha\}_{\alpha=n+1}^{n+m}$ is a parallel normal frame.

(ii) The solution of (3.14) corresponding to $\tilde{X}$ is $(\tilde{A}_1^\natural, \tilde{F}, \tilde{b})$.

(iii) The bundle morphism $P(e_\alpha(x)) = \tilde{e}_\alpha(x)$ for $n+1 \leq \alpha \leq n+m$ is a Ribaucour Transformation that covers the map $X(x) \mapsto \tilde{X}(x)$.

Next we give the geometric transformation for n-tuples in R^m of type $O(n)$ corresponding to the action of $g_{s,\pi}$ on the space of solutions of the $G_{m,n}$-system II.

10.6 Theorem. *Let $\xi = \begin{pmatrix} F \\ G \end{pmatrix}$ be a solution of the $G_{m,n}$-system (3.1), E a frame of ξ, $E(x, 0) = \begin{pmatrix} A(x) & 0 \\ 0 & B(x) \end{pmatrix}$, (F, G, B) the corresponding solution of the $G_{m,n}$-system II (3.10), and*

$$(\tilde{F}, \tilde{G}, \tilde{B^\natural}, \tilde{E^\natural}^{II}) = g_{s,\pi} \cdot (F, G, B, E^{II}), \quad \tilde{A^\natural} = g_{s,\pi} \cdot A.$$

Let e_i and $\tilde{e}_i$ denote the i-th columns of A and $\tilde{A^\natural}$ respectively. Then:

(i)

$$\frac{\partial E}{\partial \lambda}(x, 0) E^{-1}(x, 0) = \begin{pmatrix} 0 & X(x) \\ -X(x)^t & 0 \end{pmatrix},$$

$$\frac{\partial \tilde{E^\natural}}{\partial \lambda}(x, 0) \tilde{E^\natural}^{-1}(x, 0) = \begin{pmatrix} 0 & \tilde{X}(x) \\ -\tilde{X}^t(x) & 0 \end{pmatrix}$$

for some X and $\tilde{X}$.

(ii) $X = (X_1, \cdots, X_n)$ and $\tilde{X} = (\tilde{X}_1, \cdots, \tilde{X}_n)$ are n-tuples in $\mathbf{R}^m$ of type $O(n)$ such that $\{e_\alpha\}_{\alpha=n+1}^m$ and $\{\tilde{e}_\alpha\}_{\alpha=n+1}^m$ are parallel normal frame for X_j and $\tilde{X}_j$ respectively for all $1 \leq j \leq n$.

(iii) The solutions of the $G_{m,n}$-system II (3.10) corresponding to X and $\tilde{X}$ as given in Theorem 6.8 are (F, G, B) and $(\tilde{F}, \tilde{G}, \tilde{B^\natural})$ respectively.

(iv) The bundle morphism $P(e_\alpha(x)) = \tilde{e}_\alpha(x)$ for $n+1 \leq \alpha \leq m$ is a Ribaucour Transformation covering the map $X_j(x) \mapsto \tilde{X}_j(x)$ for each $1 \leq j \leq n$.

(v) There exist maps ϕ_{ij} such that $X_j + \phi_{ij} e_i = \tilde{X}_j + \phi_{ij} \tilde{e}_i$ for $1 \leq j \leq n$ and $1 \leq i \leq m$.

PROOF. It follows from Theorem 6.8, Proposition 6.10, Corollary 6.11 and formula (9.21) that

$$\begin{aligned}\begin{pmatrix} 0 & \tilde{X} \\ -\tilde{X}^t & 0 \end{pmatrix} &= \frac{\partial \tilde{E}^\natural}{\partial \lambda}(x,0)\tilde{E}^{\natural\,-1}(x,0) \\ &= \frac{\partial E}{\partial \lambda}(x,0)E(x,0)^{-1} + \frac{2}{s}E(x,0)\begin{pmatrix} 0 & \hat{W}\hat{Z}^t \\ -\hat{Z}\hat{W}^t & 0 \end{pmatrix}E(x,0)^{-1} \\ &= \frac{\partial E}{\partial \lambda}(x,0)E(x,0)^{-1} + \frac{2}{s}\begin{pmatrix} 0 & A\hat{W}\hat{Z}^tB^{-1} \\ -B\hat{Z}\hat{W}^tA^{-1} & 0 \end{pmatrix} \\ &= \begin{pmatrix} 0 & X \\ -X^t & 0 \end{pmatrix} + \frac{2}{s}\begin{pmatrix} 0 & A\hat{W}\hat{Z}^tB^{-1} \\ -B\hat{Z}\hat{W}^tA^{-1} & 0 \end{pmatrix}.\end{aligned}$$

So

$$\tilde{X} = X + \frac{2}{s}A\hat{W}\hat{Z}^tB^t.$$

Let $\eta = \sum_{j=1}^m \hat{w}_j e_j$. Then

$$\tilde{X}_i = X_i + \frac{2}{s}\sum_{j=1}^n \hat{z}_j b_{ji}\,\eta.$$

By (9.22), we get $\tilde{e}_i = e_i - 2\hat{w}_i\eta$. Hence for each $1 \le l \le m$ and $1 \le i \le n$ we have

$$X_j + \phi_{ij}e_i = \tilde{X}_j + \phi_{ij}\tilde{e}_i, \quad \text{where } \phi_{ij} = \frac{\sum_{l=1}^n \tilde{z}_l b_{lj}}{s\tilde{w}_i}.$$

This finishes the proof. ∎

10.7 Example. Recall that when $n = 2$ and $m = 3$, the $G_{3,2}$- system II (3.10) is (6.13), which is the Gauss-Codazzi equation for surfaces in R^3 parametrized by spherical line of curvature coordinates. Let (u, r_1, r_2) be a solution of (6.13), and (F, G, B) defined by (6.14) the corresponding solution of (3.10). Let $\hat{W}, \hat{Z}$ as in Theorem 9.3 for the solution $\xi = \begin{pmatrix} F \\ G \end{pmatrix}$ and E a frame, and $(\tilde{F}, \tilde{G}, \tilde{B}^\natural) = g_{s,\pi} \cdot (F, G, B)$. Then

$$\tilde{F} = \begin{pmatrix} 0 & \tilde{u}_x \\ -\tilde{u}_y & 0 \end{pmatrix}, \quad \tilde{G} = (\tilde{r}_1, \tilde{r}_2), \quad \tilde{B}^\natural = \begin{pmatrix} \cos\tilde{u} & \sin\tilde{u} \\ \sin\tilde{u} & -\cos\tilde{u} \end{pmatrix}$$

for some solution $(\tilde{u}, \tilde{r}_1, \tilde{r}_2)$ of (6.13). To see this, we write

$$\hat{Z}^t = (\hat{z}_1, \hat{z}_2) = (-\sin\alpha, \cos\alpha)$$

for some function α. It follows from (9.22) that

$$\begin{aligned}\tilde{B}^\natural = B(I - 2\hat{Z}\hat{Z}^t) &= \begin{pmatrix} \cos u & \sin u \\ -\sin u & \cos u \end{pmatrix}\begin{pmatrix} \cos 2\alpha & \sin 2\alpha \\ \sin 2\alpha & -\cos 2\alpha \end{pmatrix} \\ &= \begin{pmatrix} \cos(-u+2\alpha) & \sin(-u+2\alpha) \\ \sin(-u+2\alpha) & -\cos(-u+2\alpha) \end{pmatrix}.\end{aligned}$$

Let $X = (X_1, X_2)$ and $\tilde{X} = (\tilde{X}_1, \tilde{X}_2)$ be the 2-tuples in R^2 of type $O(2)$ corresponding to (F, G, B) and $(\tilde{F}, \tilde{G}, \tilde{B}^\natural)$ as in Theorem 10.6. Then the first and second fundamental forms of $\tilde{X}_1$ and $\tilde{X}_2$ are given by

$$\begin{aligned} I_1 &= \cos^2(-u+2\alpha)dx^2 + \sin^2(-u+2\alpha)dy^2, \\ II_1 &= -\tilde{r}_1\cos(-u+2\alpha)dx^2 - \tilde{r}_2\sin(-u+2\alpha)dy^2, \\ I_2 &= \sin^2(-u+2\alpha)dx^2 + \cos^2(-u+2\alpha)dy^2, \\ II_2 &= -\tilde{r}_1\sin(-u+2\alpha)dx^2 + \tilde{r}_2\cos(-u+2\alpha)dy^2, \quad \text{where} \\ \tilde{r}_i &= r_i - 2s\hat{w}_3\hat{z}_i = r_i - \frac{2s\tilde{w}_3\tilde{z}_i}{\tilde{z}_1^2 + \tilde{z}_1^2}. \end{aligned}$$

10.8 Example. Let $(u, r_1, r_2) = (0,0,0)$ be the trivial solution of (6.13). It is easy to see that the 2-tuple in R^3 of type $O(2)$ corresponding to the trivial solution is

$$X = \begin{pmatrix} -x & 0 \\ 0 & -y \\ 0 & 0 \end{pmatrix}$$

and

$$E(x,y,\lambda) = \begin{pmatrix} \cos\lambda x & 0 & 0 & -\sin\lambda x & 0 \\ 0 & \cos\lambda y & 0 & 0 & -\sin\lambda y \\ 0 & 0 & 1 & 0 & 0 \\ \sin\lambda x & 0 & 0 & \cos\lambda x & 0 \\ 0 & \sin\lambda y & 0 & 0 & \cos\lambda y \end{pmatrix}$$

is a frame for the trivial solution.

Below we write down explicitly the 2-tuple in R^3 of type $O(2)$ constructed by applying the Ribaucour transformation to the trivial solution. It follows from Theorems 10.6 (i) and 9.3 that

$$\begin{pmatrix} \tilde{w}_1 \\ \tilde{w}_2 \\ \tilde{w}_3 \\ \tilde{z}_1 \\ \tilde{z}_2 \end{pmatrix} = \begin{pmatrix} \cosh(sx)w_1 + \sinh(sx)z_1 \\ \cosh(sy)w_2 + \sinh(sy)z_2 \\ w_3 \\ \cosh(sx)z_1 + \sinh(sx)w_1 \\ \cosh(sy)z_2 + \sinh(sy)w_2 \end{pmatrix}. \tag{10.5}$$

Use Theorem 10.6 and a direct computation to get:

(i) If $|z_1| \geq |w_1|$ and $|z_2| \geq |w_2|$, then

$$\tilde{X}_1 = -\begin{pmatrix} x \\ 0 \\ 0 \end{pmatrix} + \frac{2a\cosh(sx)}{s(a^2\cosh^2(sx) + b^2\cosh^2(sy))}\begin{pmatrix} a\sinh(sx) \\ b\sinh(sy) \\ w_3 \end{pmatrix},$$

$$\tilde{X}_2 = -\begin{pmatrix} x \\ 0 \\ 0 \end{pmatrix} + \frac{2b\cosh(sy)}{s(a^2\cosh^2(sx) + b^2\cosh^2(sy))}\begin{pmatrix} a\sinh(sx) \\ b\sinh(sy) \\ w_3 \end{pmatrix},$$

where $a = \sqrt{z_1^2 - w_1^2}$, $b = \sqrt{z_2^2 - w_2^2}$, and $a^2 + b^2 = w_3^2 < 1$. See Figure 1.

(ii) If $|w_1| \geq |z_1|$ and $|z_2| \geq |w_2|$, then

$$\tilde{X}_1 = -\begin{pmatrix} x \\ 0 \\ 0 \end{pmatrix} + \frac{2a\sinh(sx)}{s(a^2\sinh^2(sx) + b^2\cosh^2(sy))}\begin{pmatrix} a\cosh(sx) \\ b\sinh(sy) \\ w_3 \end{pmatrix},$$

$$\tilde{X}_2 = -\begin{pmatrix} x \\ 0 \\ 0 \end{pmatrix} + \frac{2b\cosh(sy)}{s(a^2\sinh^2(sx) + b^2\cosh^2(sy))}\begin{pmatrix} a\cosh(sx) \\ b\sinh(sy) \\ w_3 \end{pmatrix},$$

where $a = \sqrt{w_1^2 - z_1^2}$, $b = \sqrt{z_2^2 - w_2^2}$, and $0 \leq b^2 - a^2 = w_3^2 < 1$. See Figure 2.

(iii) The first and second fundamental forms of $\tilde{X}_1$ are

$$I_1 = \cos^2(2\alpha)dx^2 + \sin^2(2\alpha)dy^2,$$
$$II_1 = -\frac{2sw_3}{\tau}\sin(\alpha)\cos(2\alpha)dx^2 + \frac{2sw_3}{\tau}\cos(\alpha)\sin(2\alpha)dy^2,$$

where $\tilde{z}_1, \tilde{z}_2$ are given by (10.5), $\tan\alpha = -\tilde{z}_1/\tilde{z}_2$, and

$$\tau = \{(\cosh(sx)z_1 - \sinh(sx)w_1)^2 + (\cosh(sy)z_2 - \sinh(sy)w_2)^2\}^{\frac{1}{2}}.$$

We have seen in Example 6.16 that
(i) solutions of SGE correspond to $K = -1$ surfaces of $\boldsymbol{R}^3$ up to rigid motion,
(ii) $(u, \sin u, -\cos u)$ is a solution of (6.13) if and only if u is a solution of the SGE,
(iii) if (X_1, X_2) is the 2-tuple corresponding to $(u, \sin u, -\cos u)$, then X_1 has Gaussian curvature -1, u is the corresponding solution of the SGE, and X_2 is the unit normal of X_1.

Below we give a condition on s, π so that the action of $g_{s,\pi}$ on (X, e_3) also represents a 2-tuple in R^3 of type $O(2)$ corresponding to a $K = -1$ surface.

10.9 Corollary. *Let $(u, \sin u, -\cos u)$ be a solution of (6.13), $Z = (z_1, z_2)^t$, $W = (w_1, w_2, w_3)^t$ real constant unit vectors, and constant s such that*

$$sw_3 = z_1 \sin u(0,0) - z_2 \cos u(0,0).$$

Let π denote the Hermitian projection onto $(w_1, w_2, w_3, iz_1, iz_2)^t$. Then the new solution $g_{s,\pi} \cdot (u, \sin u, -\cos u)$ of (6.13) is of the form $(\tilde{u}, \sin\tilde{u}, -\cos\tilde{u})$ for some $\tilde{u}$.

PROOF. Let $\tilde{Z}, \tilde{W}, \hat{Z}, \hat{W}$ be as in Theorem 9.3 for the solution $\xi = \begin{pmatrix} F \\ G \end{pmatrix}$ of the $G_{3,2}$-system, where F and G are defined by (6.14). Since $(\tilde{W}, \tilde{Z})$ satisfies (9.19), we have

$$\begin{cases} d\tilde{w}_3 = \sin u\ \tilde{w}_1 dx_1 - \cos u\ \tilde{w}_2 dx_2, \\ d\tilde{z}_1 = -\tilde{z}_1 du + s\tilde{w}_1 dx_1, \\ d\tilde{z}_2 = \tilde{z}_1 du + s\tilde{w}_2 dx_2. \end{cases}$$

Thus $sd\tilde{w}_3 = \sin u\ (d\tilde{z}_1 + \tilde{z}_2 du) - \cos u\ (d\tilde{z}_2 - \tilde{z}_1 du) = d(\sin u\ \tilde{z}_1 - \cos u\ \tilde{z}_2)$. It follows that $s\hat{w}_3 = \sin u\hat{z}_1 - \cos u\hat{z}_2$. ∎

10.10 Example. $u = 0$ is the trivial solution of the SGE. It follows from Example 6.16 that $(0, 0, -1)$ is a solution of (6.13). Use Corollary 10.9 and Example 10.7 to compute $g_{s,\pi} \cdot (0, 0, -1)$ to get three parameter family of solutions of the SGE. Some of these examples (see Figure 3) are:

(1) If $0 < s = \sin c \leq 1$, then

$$\tilde{u} = 2\tan^{-1}\left(\frac{\sin c \sin(y\cos c)}{\cos c\cosh(x\sin c)}\right). \tag{10.6}$$

This is the breather solution for the SGE, and the corresponding curvature -1 surface in R^3 is given by

$$\tilde{X} = -\begin{pmatrix} x \\ 0 \\ 0 \end{pmatrix} + r\begin{pmatrix} \sinh(x\sin c) \\ \cos y\cos(y\cos c) + \sec c\sin y\sin(y\cos c) \\ \sin y\cos(y\cos c) - \sec c\cos y\sin(y\cos c) \end{pmatrix},$$

where

$$r = \frac{2\cosh(x\sin c)}{\sin c\{\cosh^2(x\sin c) + \tan^2 c\sin^2(y\cos c)\}}.$$

(2) If $s = \cosh c > 1$, then

$$\tilde{u} = 2\tan^{-1}\left(\frac{\cosh c\cosh(y\sinh c)}{\sinh c\sinh(x\cosh c)}\right),$$

and the corresponding -1 curvature surface in R^3 is

$$\tilde{X} = -\begin{pmatrix} x \\ 0 \\ 0 \end{pmatrix} + r\begin{pmatrix} \cosh(x\cosh c) \\ \cos y\sinh(y\sinh c) + \mathrm{sech} c\sin y\cosh(y\sinh c) \\ \sin y\sinh(y\sinh c) - \mathrm{sech} c\cos y\cosh(y\sinh c) \end{pmatrix},$$

where

$$r = \frac{2\sinh(x\cosh c)}{\cosh c\{\sinh^2(x\cosh c) + \coth^2 c\ \sinh^2(y\sinh c)\}}.$$

11. Loop group actions for $G^1_{m,n}$-Systems

In this section, we construct the action of a rational map with two simple poles on the space of solutions of the $G^1_{m,n}$-system explicitly. Since the calculation and proofs are similar to those for the $G_{m,n}$-system in section 9, we only state the results.

The $G^1_{m,n}$*-reality condition* is

$$\begin{cases} \overline{g(\bar{\lambda})} = g(\lambda), \\ I_{m,n+1} g(\lambda) I_{m,n+1} = g(-\lambda), \\ g(\lambda)^t I_{n+m,1} g(\lambda)^t = I_{n+m,1}. \end{cases} \tag{11.1}$$

Let

$$\begin{aligned} G_+ =&\{g : C \to GL_C(n+m+1) \,|\, g \text{ is holomorphic, and satisfies (11.1)}\}, \\ G_- =&\{g : S^2 \to GL_C(n+m+1) \,|\, g \text{ is meromorphic, } g(\infty) = I, \text{ and} \\ &\text{satisfies (11.1)}\}. \end{aligned}$$

Let $\boldsymbol{C}^{n+m+1}$ be equipped with the bi-linear form:

$$\langle u,\ v \rangle_1 = \sum_{i=1}^{n+m} \bar{u}_i v_i - \bar{u}_{n+m+1} v_{n+m+1}.$$

Let $W = (w_1, \ldots, w_m)^t$ and $Z = (z_1, \ldots, z_{n+1})^t$ be unit vectors in R^m and the Lorentz space $R^{n,1}$ respectively, and π the orthogonal projection of $\boldsymbol{C}^{n+m+1}$ onto the span of $\begin{pmatrix} W \\ iZ \end{pmatrix}$ with respect to $\langle\ ,\ \rangle_1$. So

$$\pi = \frac{1}{2} \begin{pmatrix} WW^t & -iWZ^t \\ iZW^t & ZZ^t \end{pmatrix} I_{n+m,1}. \tag{11.2}$$

Since Z, W are real vectors, $\pi\bar{\pi} = \bar{\pi}\pi = 0$. Given non-zero $s \in R$, define

$$\begin{aligned} q_{s,\pi}(\lambda) = &\left(\pi + \frac{\lambda - is}{\lambda + is}(I - \pi)\right)\left(\bar{\pi} + \frac{\lambda + is}{\lambda - is}(I - \bar{\pi})\right) = \\ &\frac{1}{\lambda^2 + s^2}\left(\lambda^2 I + 2s\lambda \begin{pmatrix} 0 & WZ^t J \\ -ZW^t & 0 \end{pmatrix} + s^2 \begin{pmatrix} I - 2WW^t & 0 \\ 0 & I - 2ZZ^t J \end{pmatrix}\right), \end{aligned} \tag{11.3}$$

where $J = I_{n,1} = \text{diag}(1, \cdots, 1, -1)$. A direct computation implies that $q_{s,\pi}$ satisfies the reality condition (11.1). So $q_{s,\pi} \in G_-$.

11.1 Theorem. *Let ξ be a solution of the $G^1_{m,n}$-system (4.2), θ_λ its Lax connection as in (4.1), and E a frame for ξ. Let W and Z be unit vectors in R^m and $R^{n,1}$,*

$$\begin{pmatrix} \tilde{W} \\ i\tilde{Z} \end{pmatrix}(x) = E(x, -is)^{-1} \begin{pmatrix} W \\ iZ \end{pmatrix}, \tag{11.4}$$

and $\tilde{\pi}(x)$ the orthogonal projection onto the span of $\begin{pmatrix} \tilde{W} \\ i\tilde{Z} \end{pmatrix}(x)$ with respect to $\langle\,,\,\rangle_1$. Let $0 \neq s \in R$ be a constant, π a projection onto $\mathbf{C}\begin{pmatrix} W \\ iZ \end{pmatrix}$, and $q_{s,\pi}$ defined by (11.3). Set

$$\tilde{E}(x,\lambda) = q_{s,\pi}(\lambda)E(x,\lambda)q^{-1}_{s,\tilde{\pi}(x)}(\lambda),$$
$$\tilde{\xi} = \xi - 2s(\hat{W}\hat{Z}^t J)_*,$$

where $\hat{W}(x) = \frac{1}{||\tilde{W}(x)||_m}\tilde{W}(x)$ and $\hat{Z}(x) = \frac{1}{||\tilde{Z}(x)||_{n,1}}\tilde{Z}(x)$, and $(y_*)_{ij} = y_{ij}$ for $i \neq j$ and $(y_*)_{ii} = 0$ for all i. Let

$$\tilde{E^\natural}(x,\lambda) = E(x,\lambda)q^{-1}_{s,\tilde{\pi}(x)}(\lambda).$$

Then
(i) $\tilde{\xi}$ is a new solution of (4.2),
(ii) $\tilde{E}^\natural$ is a frame for $\tilde{\xi}$,
(iii) $E(x,0) = \begin{pmatrix} A(x) & 0 \\ 0 & B(x) \end{pmatrix}$ and $\tilde{E}^\natural(x,0) = \begin{pmatrix} \tilde{A}^\natural(x) & 0 \\ 0 & \tilde{B}^\natural(x) \end{pmatrix}$ for some A, B, $\tilde{A}^\natural, \tilde{B}^\natural$, and

$$\begin{cases} \tilde{A}^\natural = A(I - 2\hat{W}\hat{W}^t), \\ \tilde{B}^\natural = B(I - 2\hat{Z}\hat{Z}^t)I_{n,1}. \end{cases} \tag{11.5}$$

When $m = n$, Theorem 11.1 (i) was proved in [Zh].

Write $\xi = \begin{pmatrix} F \\ G \end{pmatrix}$, $F = (f_{ij})$, $G = (g_{ij})$, $\tilde{F} = (\tilde{f}_{ij})$, $\tilde{G} = (\tilde{g}_{ij})$, and $\tilde{\xi} = \begin{pmatrix} \tilde{F} \\ \tilde{G} \end{pmatrix}$. Then formula for $\tilde{\xi}$ in the above theorem is

$$\begin{cases} \tilde{f}_{ij} = f_{ij} - 2s\hat{w}_i\hat{z}_j\epsilon_j, \\ \tilde{g}_{ij} = g_{ij} - 2s\hat{w}_{n+1+i}\hat{z}_j\epsilon_j, \end{cases} \tag{11.6}$$

where $\epsilon_j = 1$ for $1 \leq j \leq n$ and $\epsilon_{n+1} = -1$.

Differentiate (11.4) to get the ODE for $(\tilde{W}, \tilde{Z})$:

11.2 Corollary. *Let $\xi = \begin{pmatrix} F \\ G \end{pmatrix}$ be a solution of the $G^1_{m,n}$-system (4.2), E a frame of ξ, and $(\tilde{W}, \tilde{Z})$ be as in Theorem 11.1. Then $(\tilde{W}, \tilde{Z})$ is a solution of the following system*

$$\begin{cases} (\tilde{w}_i)_{x_i} = -\sum_{j\neq i}^{n+1} f_{ji}\tilde{w}_j - \sum_{j=1}^{m-n-1} g_{ji}\tilde{w}_{j+n+1} + s\tilde{z}_i\epsilon_i, \quad i \leq n+1 \\ (\tilde{w}_i)_{x_j} = f_{ij}\tilde{w}_j, \quad i \leq n+1, j \neq i, \\ (\tilde{w}_i)_{x_j} = g_{ij}\tilde{w}_j, \quad i > n+1, \\ (\tilde{z}_i)_{x_j} = f_{ji}\tilde{z}_j\epsilon_i\epsilon_j, \quad j \neq i, \\ (\tilde{z}_i)_{x_i} = -\sum_{j\neq i}^{n+1} f_{ij}\tilde{z}_j + s\tilde{w}_i. \end{cases} \tag{11.7}$$

11.3 Corollary. *Let $\xi = \begin{pmatrix} F \\ G \end{pmatrix}$ be a solution of the $G^1_{m,n}$-system (4.2). Then system (11.7) is always solvable. Moreover, if $(\tilde{W}, \tilde{Z})$ is a solution of (11.7) such that $\tilde{W}(0)$ and $\tilde{Z}(0)$ are unit vectors in R^m and $R^{n,1}$ respectively, then $\begin{pmatrix} \tilde{F} \\ \tilde{G} \end{pmatrix}$ is a solution of (4.2), where $\tilde{F}$ and $\tilde{G}$ are defined by (11.6).*

Write

$$A = (A_1, A_2), \quad \tilde{A}^\natural = (\tilde{A}^\natural_1, \tilde{A}^\natural_2)$$

with $A_1, \tilde{A}^\natural_1 \in \mathcal{M}_{m\times(n+1)}$ and $A_2, \tilde{A}^\natural_2 \in \mathcal{M}_{m\times(m-n-1)}$. Then (A_1, F) and $(\tilde{A}^\natural_1, \tilde{F})$ are solutions of the $G^1_{m,n}$-system I (4.4), and (F, G, B) and $(\tilde{F}, \tilde{G}, \tilde{B}^\natural)$ are solutions of the $G^1_{m,n}$-system II (4.6), and their frames are related by

$$\begin{aligned}
\tilde{E}^\natural(x,\lambda) &= E(x,\lambda) q_{s,\tilde{\pi}(x)}(\lambda)^{-1}, \\
\tilde{E^\natural}^I(x,\lambda) &= E^I(x,\lambda)\left(I - \frac{2}{\lambda^2+s^2}\begin{pmatrix} \lambda A\hat{W} \\ s\hat{Z} \end{pmatrix}\begin{pmatrix} \lambda \hat{W}^t A^t & s\hat{Z}^t J \end{pmatrix}\right), \\
\tilde{E^\natural}^{II}(x,\lambda) &= E^{II}(x,\lambda)\left(I - \frac{2}{\lambda^2+s^2}\begin{pmatrix} s^2\hat{W}\hat{W}^t & -s\lambda\hat{W}\hat{Z}^t B^t J \\ -s\lambda B\hat{Z}\hat{W}^t & \lambda^2 B\hat{Z}\hat{Z}^t B^t J \end{pmatrix}\right).
\end{aligned}$$

For the partial $G^1_{m,n}$-system we have

11.4 Theorem. *Let $(F, G, b) : R^n \to gl_*(n) \times \mathcal{M}_{(m-n)\times n} \times \mathcal{M}_{n\times 1}$ be a solution of the partial $G^1_{m,n}$-system (4.8), τ_λ its Lax connection as in (4.7), and E a frame for (F, G, b). Let W and Z be unit vectors in R^m and $R^{n,1}$ respectively,*

$$\begin{pmatrix} \tilde{W} \\ i\tilde{Z} \end{pmatrix}(x) = E(x, -is)^{-1}\begin{pmatrix} W \\ iZ \end{pmatrix},$$

and $\tilde{\pi}(x)$ the orthogonal projection onto the span of $\begin{pmatrix} \tilde{W} \\ i\tilde{Z} \end{pmatrix}(x)$ with respect to the bilinear form $\langle\ ,\ \rangle_1$. Let $0 \neq s \in R$ be a real number, π the projection onto $\mathbf{C}\begin{pmatrix} W \\ iZ \end{pmatrix}$, and $q_{s,\pi}$ as defined in (11.3). Let

$$\tilde{E}(x,\lambda) = q_{s,\pi}(\lambda) E(x,\lambda) q^{-1}_{s,\tilde{\pi}(x)}(\lambda),$$

$$\begin{cases} \tilde{f}_{ij} = f_{ij} - 2s\hat{w}_i\hat{z}_j, \\ \tilde{g}_{ij} = g_{ij} - 2s\hat{w}_{n+i}\hat{z}_j, \\ \tilde{b}_i = b_i + 2s\hat{w}_i\hat{z}_{n+1}, \end{cases} \tag{11.8}$$

where $\hat{W} = \frac{1}{||\tilde{W}||_m}\tilde{W}$, $\hat{Z} = \frac{1}{||\tilde{Z}||_{n,1}}\tilde{Z}$ and $\hat{w}_i$ and $\hat{z}_i$ are the i-th coordinate of $\hat{W}, \hat{Z}$ respectively. Then $(\tilde{F}, \tilde{G}, \tilde{b})$ is a solution of (4.8) and $\tilde{E}$ is its frame, where $\tilde{F} = (\tilde{f}_{ij})$, $\tilde{G} = (\tilde{g}_{ij})$, and $\tilde{b} = (\tilde{b}_i)$.

11.5 Corollary. *Let* (F, G, b) *be a solution of the partial* $G^1_{m,n}$*-system (4.8), E a frame of* (F, G, b)*, and* $\begin{pmatrix} \tilde{W} \\ i\tilde{Z} \end{pmatrix}$ *as in Theorem 11.4. Then* $(\tilde{W}, \tilde{Z})$ *is a solution of*

$$\begin{cases} (\tilde{w}_i)_{x_i} = -\sum_{j\neq i}^n f_{ji}\tilde{w}_j - \sum_{j=1}^{m-n} g_{ji}\tilde{w}_{j+n} + s\tilde{z}_i & i \leq n\,, \\ (\tilde{w}_i)_{x_j} = f_{ij}\tilde{w}_j & i \leq n, j \neq i\,, \\ (\tilde{w}_i)_{x_j} = g_{ij}\tilde{w}_j & i > n\,, \\ (\tilde{z}_i)_{x_j} = f_{ji}\tilde{z}_j & j \neq i\,, i \leq n \\ (\tilde{z}_i)_{x_i} = -\sum_{j\neq i}^n f_{ij}\tilde{z}_j - b_i\tilde{z}_{n+1} + s\tilde{w}_i, \\ (\tilde{z}_{n+1})_{x_j} = -b_j\tilde{z}_j. \end{cases} \tag{11.9}$$

Conversely, if (F, G, b) *is a solution of system (4.8), then system (11.9) is solvable. Moreover, if* $(\tilde{W}, \tilde{Z})$ *is a solution of (11.9) such that* $\tilde{W}(0)$ *and* $\tilde{Z}(0)$ *are unit vectors in* R^m *and* $R^{n,1}$ *respectively, then* $(\tilde{F}, \tilde{G}, \tilde{b})$ *is also a solution of (4.8), where* $\tilde{F}, \tilde{G}, \tilde{b}$ *are defined by formula (11.8).*

Let

$$\tilde{E}^\natural(x, \lambda) = E(x, \lambda) q^{-1}_{s,\tilde{\pi}(x)}(\lambda).$$

Then $\tilde{E}^\natural$ is also a frame of $(\tilde{F}, \tilde{G}, \tilde{b})$. Write

$$E(x, 0) = \begin{pmatrix} A & 0 \\ 0 & B \end{pmatrix} \in O(m) \times O(n, 1), \quad \tilde{E}^\natural(x, 0) = \begin{pmatrix} \tilde{A}^\natural(x) & 0 \\ 0 & \tilde{B}^\natural(x) \end{pmatrix}.$$

Then

$$\begin{cases} \tilde{A}^\natural = A(I - 2\hat{W}\hat{W}^t), \\ \tilde{B}^\natural = B(I - 2\hat{Z}\hat{Z}^t J), \end{cases}$$

where $J = I_{n,1}$.

Let

$$E^I(x, \lambda) = E(x, \lambda) \begin{pmatrix} A^t & 0 \\ 0 & I_{n+1} \end{pmatrix}, \quad \tilde{E^\natural}^I(x, \lambda) = \tilde{E}(x, \lambda) \begin{pmatrix} \tilde{A}^t & 0 \\ 0 & I_{n+1} \end{pmatrix}.$$

A direct computation gives

$$\tilde{E^\natural}^I(x, \lambda) = E^I(x, \lambda) \left(I - \frac{2}{\lambda^2 + s^2} \begin{pmatrix} \lambda A\hat{W} \\ s\hat{Z} \end{pmatrix} \begin{pmatrix} \lambda \hat{W}^t A^t & s\hat{Z}^t J \end{pmatrix} \right).$$

Note that E^I and $\tilde{E^\natural}^I$ are frames of (A_1, F, b) and $(\tilde{A}^\natural_1, \tilde{F}, \tilde{b})$ respectively. We use the same type of notations as in previous sections:

$$\begin{aligned} &(\tilde{F}, \tilde{G}, \tilde{b}, \tilde{E}^\natural) = q_{s,\pi} \cdot (F, G, b, E), \\ &(\tilde{A}^\natural_1, \tilde{F}, \tilde{b}, \tilde{E^\natural}^I) = q_{s,\pi} \cdot (A_1, F, b, E), \\ &(\tilde{A}^\natural, \tilde{B}^\natural) = q_{s,\pi} \cdot (A, B). \end{aligned}$$

12. Ribaucour Transformations for $G^1_{m,n}$-Systems

In this section, we describe the corresponding geometric transformations on submanifolds associated to the action of $q_{s,\pi}$ on the space of solutions of various $G^1_{m,n}$-systems described in section 11.

12.1 Theorem. *Let X be a local isometric immersion of H^n in H^{n+m} with flat and non-degenerate normal bundle, $(x_1, \ldots, x_n)$ line of curvature coordinates, and (A_1, F, b) the solution of the $G^1_{m,n}$-system I (4.10) corresponding to X as defined in Theorem 7.3. Let $E^I(x,\lambda)$ be a frame of (A_1, F, b), and $g(x) = E^I(x,1)$. Let $0 \neq s \in R$, π the projection onto $\mathbf{C}\begin{pmatrix} W \\ iZ \end{pmatrix}$, $q_{s,\pi}$ defined by (11.3), and $\tilde{E^\natural}^I$, $\tilde{A}_i^\natural$ as in section 11. Write*

$$
\begin{aligned}
(\tilde{A}_1^\natural, \tilde{F}, \tilde{b}, \tilde{E^\natural}^I) &= q_{s,\pi} \cdot (A_1, F, b, E^I), \\
E^I(x,1) &= (e_{n+1}(x), \cdots, e_{n+m}(x), e_1(x), \cdots, e_n(x), X(x)), \\
\tilde{E^\natural}^I(x,1) &= (\tilde{e}_{n+1}(x), \cdots, \tilde{e}_{n+m}(x), \tilde{e}_1(x), \cdots, \tilde{e}_n(x), \tilde{X}(x)).
\end{aligned}
$$

Then:

(i) $\tilde{X}$ is a local isometric immersion of H^n in H^{n+m} with non-degenerate, flat normal bundle, and $\{e_\alpha\}_{\alpha=n+1}^{n+m}$ is a parallel normal frame.

(ii) The solution of (4.10) corresponding to $\tilde{X}$ given in Theorem 7.3 is $(\tilde{A}_1^\natural, \tilde{F}, \tilde{b})$.

(iii) The bundle morphism $P(e_\alpha(x)) = \tilde{e}_\alpha(x)$ for $n+1 \leq \alpha \leq n+m$ is a Ribaucour transformation covering the map $X(x) \mapsto \tilde{X}(x)$.

Recall that if v_0 is a null vector in $R^{2n,1}$ and $c \neq 0$, then by Proposition 7.1 that

$$N_{v_0,c} = \{x \in^{2n,1} \mid \langle x, x\rangle_1 = -1, \langle x, v_0\rangle_1 = c\}$$

is a totally umbilic hypersurface of H^{2n}, and $N_{v_0,c}$ is isometric to R^{2n-1}. Next we give a condition on s, π so that the Ribaucour transformation corresponding to $q_{s,\pi}$ for local isometric immersions of H^n in H^{2n} preserves local isometric immersions of H^n in R^{2n-1}.

12.2 Corollary. *Let v_0 be a null vector in $R^{2n,1}$, X a local isometric immersion of H^n in $N_{v_0,c} \subset H^{2n}$ with flat and non-degenerate normal bundle, and W, Z, $q_{s,\pi}$ and $\tilde{X}$ as in Theorem 12.1. If*

$$b^t(0)W = -sz_{n+1},$$

then $\tilde{X}$ also lies in some flat totally umbilic hypersurface of H^{2n}.

PROOF. By Theorem 7.3 (vi), there exists some constant vector $w \in \mathcal{M}_{n\times 1}$ such that $b = A^t w$. It follows from 2.4 that

$$
\begin{cases}
d\tilde{w} = -\sum(-FC_i + C_iF^t)\tilde{w}dx_i + s\sum \tilde{z}_iC_idx_i, \\
d\tilde{z}_{n+1} = -\sum b_i\tilde{z}_idx_i.
\end{cases}
$$

Since $A^t dA = \sum(-FC_i + C_iF^t)dx_i$, we have

$$d(b^t\tilde{W}) = w^t d(A\tilde{W}) = w^t(dA)\tilde{W} + w^t Ad\tilde{W} = b^t(s\sum \tilde{z}_i C_i dx_i) = -sd\tilde{z}_{n+1}.$$

Hence $b^t\hat{W} = -s\hat{z}_{n+1}$ for all $x \in R^n$.A straightforward calculation now implies that

$$\tilde{A^\natural}^t w = A(I - 2\hat{W}\hat{W}^t)A^t w = b + 2s\hat{z}_{n+1}\hat{W} = \tilde{b},$$

which finishes the proof. ∎

12.3 Theorem. *Let (F, G, B) be a solution of the $G^1_{m,n}$-system II (4.6), and $X = (X_1, \cdots, X_{n+1})$ the $(n+1)$-tuple in R^m of type $O(n,1)$ corresponding to (F, G, B) of (4.6) as described in Theorem 7.4. Let $q_{s,\pi}$ be the rational map defined by (11.3) for some constant s and unit vectors W in R^m and Z in $R^{n,1}$, and*

$$(\tilde{F}, \tilde{G}, \tilde{B}^\natural, \tilde{A}^\natural) = q_{s,\pi} \cdot (F, G, B, A).$$

Let

$$\tilde{X} = X + \frac{2}{s} A\hat{W}\hat{Z}^t B^t J = (\tilde{X}_1, \cdots, \tilde{X}_{n+1}), \tag{12.1}$$

where $J = I_{n,1}$. Then:

(i) $\tilde{X}$ is an $(n+1)$-tuple in R^m of type $O(n,1)$.

(ii) The solution of (4.6) corresponding to $\tilde{X}$ is $(\tilde{F}, \tilde{G}, \tilde{B}^\natural) = q_{s,\pi} \cdot (F, G, B)$.

(iii) Write $A = (e_1, \cdots, e_m)$, $\tilde{A}^\natural = (\tilde{e}_1, \cdots, \tilde{e}_m)$. Then $\{e_i\}_{i=1}^n$ and $\{\tilde{e}_i\}_{i=1}^n$ are principal curvature frames, and $\{e_\alpha\}_{\alpha=n+1}^m$ and $\{\tilde{e}_\alpha\}_{\alpha=n+1}^m$ are parallel normal frames of X_j and $\tilde{X}_j$ respectively.

(iv) The bundle morphism $P(e_\alpha(x)) = \tilde{e}_\alpha(x)$ for $n+2 \leq \alpha \leq m$ is a Ribaucour transformation covering the map $X_j(x) \mapsto \tilde{X}_j(x)$ for each $1 \leq j \leq n$.

(v) There exist smooth functions ϕ_{ij} such that

$$X_j + \phi_{ij}e_i = \tilde{X}_j + \phi_{ij}\tilde{e}_i \tag{12.2}$$

for all $1 \leq j \leq n$ and $1 \leq i \leq n$.

13. Darboux Transformations for $G^1_{m,1}$-Systems

Let $M, \tilde{M}$ be two surfaces in R^m with flat and non-degenerate normal bundle, and $P : \nu(M) \to \nu(\tilde{M})$ a Ribaucour transformation that covers $\ell : M \to \tilde{M}$. If in addition ℓ is a conformal diffeomorphism, then P is called a *Darboux transformation*. Such transformations for surfaces in R^3 were studied by Darboux and Bianchi and in R^m by Burstall [Bu]. In this section, we show that the transformation constructed in Theorem 12.3 for 2-tuples in R^3 of type $O(1,1)$ is a Darboux transformation for isothermic surfaces.

13.1 Theorem. *Let (Y_1, Y_2) be an isothermic pair in R^m corresponding to the solution (u, G) of (8.2), and $\xi = \begin{pmatrix} F \\ G \end{pmatrix}$ the corresponding solution of the $G^1_{m,1}$-system, where $F = \begin{pmatrix} 0 & u_x \\ u_y & 0 \end{pmatrix}$. Let $0 \neq s \in R$, π the projection onto $\mathbf{C}\begin{pmatrix} W \\ iZ \end{pmatrix}$, $q_{s,\pi}$ the rational map defined by (11.3), and $\hat{W}, \hat{Z}$ as in Theorem 11.1 for the solution $\xi = \begin{pmatrix} F \\ G \end{pmatrix}$ of the $G^1_{m,1}$-system. Let $\tilde{E^\natural}^{II}, \tilde{A^\natural}, \tilde{B^\natural}$ be as in section 11, and*

$$B = \begin{pmatrix} \cosh u & \sinh u \\ \sinh u & \cosh u \end{pmatrix}, \quad (\tilde{E^\natural}^{II}, \tilde{A^\natural}, \tilde{B^\natural}) = q_{s,\pi} \cdot (E^{II}, A, B).$$

Write $A = (e_1, \cdots, e_m)$ and $\tilde{A^\natural} = (\tilde{e}_1, \cdots, \tilde{e}_m)$. Set

$$\begin{aligned} \tilde{Y_1} &= Y_1 + \frac{2}{s}(\hat{z}_1 + \hat{z}_2)e^u \sum_{i=1}^m \hat{w}_i e_i, \\ \tilde{Y_2} &= Y_2 + \frac{2}{s}(\hat{z}_1 - \hat{z}_2)e^{-u} \sum_{i=1}^m \hat{w}_i e_i. \end{aligned} \tag{13.1}$$

Then
(i) $(\tilde{Y}_1, \tilde{Y}_2)$ is an isothermic pair,
(ii) $(\tilde{u}, \tilde{G})$ is the solution of (8.2) corresponding to $(\tilde{Y}_1, \tilde{Y}_2)$, where $\tilde{u} = 2\alpha - u$, $\sinh \alpha = -\hat{z}_2$ and $\tilde{G} = (\tilde{g}_{ij})$ is defined by (11.6),
(iii) the fundamental forms of $(\tilde{Y}_1, \tilde{Y}_2)$ are

$$\begin{cases} \tilde{I}_1 = e^{-2\tilde{u}}(dx^2 + dy^2), \\ \tilde{II}_1 = e^{-\tilde{u}} \sum_{j=1}^{m-2} (\tilde{g}_{j1} dx^2 - \tilde{g}_{j2} dy^2) \tilde{e}_{2+j}, \end{cases}$$
$$\begin{cases} \tilde{I}_2 = e^{2\tilde{u}}(dx^2 + dy^2), \\ \tilde{II}_2 = e^{\tilde{u}} \sum_{j=1}^{m-2} (\tilde{g}_{j1} dx^2 + \tilde{g}_{j2} dy^2) \tilde{e}_{j+2}, \end{cases}$$

(iv) the bundle morphism $P(e_\alpha(x)) = \tilde{e}_\alpha(x)$ for $3 \leq \alpha \leq m$ covering $Y_i \mapsto \tilde{Y}_i$ is a Darboux transformation for each $1 \leq i \leq 2$.

PROOF. Let $X_1 = (Y_1 + Y_2)/2$, and $X_2 = (Y_2 - Y_1)/2$. By Propositions 8.1, 8.5 and 8.6, (F, G, B) is a solution of the $G^1_{m,1}$-system II, and $X = (X_1, X_2)$ is the corresponding 2-tuple in R^m of type $O(1,1)$. Let $\tilde{X} = (\tilde{X}_1, \tilde{X}_2)$ be as in Theorem 12.3, and

$$\tilde{Y_1} = \tilde{X}_1 - \tilde{X}_2, \quad \tilde{Y_2} = \tilde{X}_1 + \tilde{X}_2.$$

It follows from (12.1) that $\tilde{Y}_1, \tilde{Y}_2$ are given by (13.1).
Since $\hat{z}_1^2 - \hat{z}_2^2 = 1$, there exists a function $\alpha : R^2 \to R$ such that

$$\hat{z}_1 = \cosh \alpha, \quad \hat{z}_2 = -\sinh \alpha.$$

Since $\tilde{u} = 2\alpha - u$,

$$e^{-\tilde{u}} = e^u \frac{\hat{z}_1 + \hat{z}_2}{\hat{z}_1 - \hat{z}_2}. \tag{13.2}$$

Use (11.5), (13.2) and a direct computation to get

$$\begin{aligned} \tilde{B}^\natural = B(I - 2\hat{Z}\hat{Z}^t J) &= \begin{pmatrix} -\cosh(2\alpha - u) & -\sinh(2\alpha - u) \\ \sinh(2\alpha - u) & \cosh(2\alpha - u) \end{pmatrix} \\ &= \begin{pmatrix} -\cosh\tilde{u} & -\sinh\tilde{u} \\ \sinh\tilde{u} & \cosh\tilde{u} \end{pmatrix}. \end{aligned} \tag{13.3}$$

So (i)-(iii) follow.

To prove (iv), we first note that the map $Y_i \mapsto \tilde{Y}_i$ is conformal because (x, y) are isothermic coordinates for Y_i and $\tilde{Y}_i$. It remains to prove that P is a Ribaucour transform. To see this, we use (12.2) to get

$$\begin{aligned} Y_1 + (\phi_{i1} - \phi_{i2})e_i &= \tilde{Y}_1 + (\phi_{i1} - \phi_{i2})\tilde{e}_i, \\ Y_2 + (\phi_{i1} + \phi_{i2})e_i &= \tilde{Y}_2 + (\phi_{i1} + \phi_{i2})\tilde{e}_i \end{aligned}$$

for each $1 \le i \le m$. So P is a Darboux transformation. ∎

Let (Y_1, Y_2) be an isothermic pair in R^3 corresponding to the solution (u, G) of (8.2), and $(\tilde{Y}_1, \tilde{Y}_2)$ as in Theorem 13.1. Write $G = (g_1, g_2)^t$. Then the mean curvature for Y_i and $\tilde{Y}_i$ are given as follows:

$$\begin{aligned} H_1 &= \frac{-g_1 - g_2}{e^u} \\ H_2 &= \frac{-g_1 + g_2}{e^{-u}} \\ \tilde{H}_1 &= \frac{\tilde{g}_1 - \tilde{g}_2}{e^{-\tilde{u}}} = \frac{\hat{z}_1 - \hat{z}_2}{e^u(\hat{z}_1 + \hat{z}_2)}\{g_1 - g_2 - 2s\hat{w}_3(\hat{z}_1 + \hat{z}_2)\} \\ \tilde{H}_2 &= \frac{\tilde{g}_1 + \tilde{g}_2}{e^{\tilde{u}}} = \frac{e^u(\hat{z}_1 + \hat{z}_2)}{\hat{z}_1 - \hat{z}_2}\{g_1 + g_2 - 2s\hat{w}_3(\hat{z}_1 - \hat{z}_2)\}. \end{aligned} \tag{13.4}$$

13.2 Example. A plane in R^3 is the isothermic surface corresponding to the trivial solution $(0, 0, 0)$ of (8.2). Let $W = (w_1, w_2, w_3)^t$, $Z = (z_1, z_2)^t$ be unit vectors in R^3 and $R^{1,1}$ respectively. Use Theorem 11.1 and a direct computation to get

$$\begin{pmatrix} \tilde{w}_1 \\ \tilde{w}_2 \\ \tilde{w}_3 \\ \tilde{z}_1 \\ \tilde{z}_2 \end{pmatrix}(x, y) = \begin{pmatrix} w_1 \cosh sx + z_1 \sinh sx \\ w_2 \cos sy + z_2 \sin sy \\ w_3 \\ z_1 \cosh sx + w_1 \sinh sx \\ -w_2 \sin sy + z_2 \cos sy \end{pmatrix}.$$

Then the isothermic pair $(\tilde{Y}_1, \tilde{Y}_2)$ corresponding to $(\tilde{u}, \tilde{g}_1, \tilde{g}_2) = q_{s,\pi} \cdot (0,0,0)$ is given by

$$\tilde{Y}_1 = \begin{pmatrix} -x \\ y \\ 0 \end{pmatrix} + \frac{2}{s\{\cosh\gamma\cosh(sx) - \cos(sy)\}} \begin{pmatrix} \cosh\gamma\sinh(sx) \\ \sin(sy) \\ -\sinh\gamma \end{pmatrix},$$

$$\tilde{Y}_2 = \begin{pmatrix} -x \\ -y \\ 0 \end{pmatrix} + \frac{2}{s\{\cosh\gamma\cosh(sx) + \cos(sy)\}} \begin{pmatrix} \cosh\gamma\sinh(sx) \\ \sin(sy) \\ -\sinh\gamma \end{pmatrix},$$

where $\cosh\gamma = \frac{\sqrt{z_1^2 - w_1^2}}{\sqrt{z_2^2 + w_2^2}}$ is a constant parameter. See Figure 4.

The first and second fundamental forms of $\tilde{Y}_2$ are given by $\tilde{I} = e^{2\tilde{u}}(dx^2 + dy^2)$ and $\tilde{II} = e^{2\tilde{u}}(\tilde{k}_1 dx^2 + \tilde{k}_2 dy^2)$ where

$$\begin{aligned} e^{\tilde{u}} &= \frac{\cosh\gamma\cosh(sx) - \cos(sy)}{\cosh\gamma\cosh(sx) + \cos(sy)}, \\ \tilde{k}_1 &= \frac{2s\sinh\gamma\cosh\gamma\cosh(sx)}{(\cosh\gamma\cosh(sx) - \cos(sy))^2}, \\ \tilde{k}_2 &= \frac{2s\sinh\gamma\cos(sy)}{(\cosh\gamma\cosh(sx) - \cos(sy))^2}. \end{aligned}$$

Suppose $Y_1(x,y)$ is an immersion of a surface in R^3 with constant mean curvature $2c$ and no umbilical points, and (x,y) is the canonical isothermic coordinates. It follows from Proposition 8.7 that (u, g_1, g_2) is a solution of (8.2), where

$$g_1 = e^{-u} + ce^u, \quad g_2 = -e^{-u} + ce^u.$$

In particular, Y_1 is isothermic. It follows from Theorem 7.4, Propositions 8.1 and 8.6 that there exists an isothermic surface Y_2 such that (Y_1, Y_2) is an isothermic pair corresponding to (u, g_1, g_2), and the principal curvatures of Y_2 are

$$-1 - ce^{2u}, \quad -1 + ce^{2u}.$$

So the mean curvature of Y_2 is -2. If Y_1 is minimal, then Y_2 is totally umbilic with mean curvature -2. Hence in this case, Y_2 is a standard unit sphere. In fact, $Y_2 = e_3$ is the standard sphere parametrized by the isothermic coordinates of Y_1.

Below we give a condition on s, π so that the induced action of $q_{s,\pi}$ on the space of isothermic surfaces preserves the subset of constant mean curvature surfaces parametrized by canonical isothermic coordinates.

13.3 Corollary. *Let (Y_1, Y_2) be a isothermic pair in R^3 such that Y_1 has constant mean curvature $2c$, no umbilical points, and is parametrized by the canonical isothermic coordinates. Let $s \neq 0$ be a constant, W, Z constant unit vectors in R^m and $R^{1,1}$ respectively, and $\tilde{Y} = (\tilde{Y}_1, \tilde{Y}_2) = q_{s,\pi} \cdot (Y_1, Y_2)$ as in Theorem 13.1. If*

$$sw_3 = ce^{u(0,0)}(z_1 + z_2) + e^{-u(0,0)}(z_1 - z_2),$$

then $\tilde{Y}_1$ has constant mean curvature $2c$ and $\tilde{Y}_2$ has mean curvature -2.

PROOF. By (11.8), we have

$$\tilde{g}_1 = g_1 - 2s\hat{w}_3\hat{z}_1, \quad \tilde{g}_2 = g_2 + 2s\hat{w}_3\hat{z}_2.$$

Write $\hat{z}_1 = \cosh\alpha$ and $\hat{z}_2 = -\sinh\alpha$ for some function α. It follows from the differential equation (11.7) for $\tilde{W}, \tilde{Z}$ that the derivative of

$$s\tilde{w}_3 - ce^u(\tilde{z}_1 + \tilde{z}_2) - e^{-u}(\tilde{z}_1 - \tilde{z}_2) \tag{13.5}$$

is zero. Since $\tilde{W}(0) = W$ and $\tilde{Z}(0) = Z$, (13.5) is identically zero on R^2. Use Theorem 13.1, formulas (13.3), (13.5), and a direct computation to prove that the mean curvature of $\tilde{Y}_1$ and $\tilde{Y}_2$ are $2c$ and -2 respectively. ∎

Similar computation gives

13.4 Corollary. *Let Y_1 be a minimal surface without umbilic points in R^3 parametrized by the canonical isothermic coordinates, and $\tilde{Y}_1$ as in Corollary 13.3. If $sw_3 = e^{-u(0,0)}(z_1 - z_2)$, then $\tilde{Y}_1$ is minimal.*

13.5 Example.

Let $Y_1 = \begin{pmatrix} -x \\ \sin y \\ 1-\cos y \end{pmatrix}$ and $Y_2 = \begin{pmatrix} -x \\ -\sin y \\ -1+\cos y \end{pmatrix}$. This pair of cylinders (Y_1, Y_2) is an isothermic pair with $(u, g_1, g_2) = (0, 0, -1)$ as the corresponding solution to (8.2). Using Theorem 11.1 and putting $s = \sinh c$, we obtain

$$\begin{pmatrix} \tilde{w}_1 \\ \tilde{w}_2 \\ \tilde{w}_3 \\ \tilde{z}_1 \\ \tilde{z}_2 \end{pmatrix} = \begin{pmatrix} w_1\cosh(x\sinh c) + z_1\sinh(x\sinh c) \\ w_2\cos(y\cosh c) + w_3\cosh c\sin(y\cosh c) \\ -w_2\operatorname{sech} c\ \sin(y\cosh c) + w_3\cos(y\cosh c) \\ z_1\cosh(x\sinh c) + w_1\sinh(x\sinh c) \\ \sinh c\ w_2\operatorname{sech} c\ \sin(y\cosh c) - \sinh c\ w_3\cos(y\cosh c) \end{pmatrix}.$$

Let $(\tilde{Y}_1, \tilde{Y}_2)$ be the isothermic pair with constant mean curvatures corresponding to $(\tilde{u}, \tilde{g}_1, \tilde{g}_2) = q_{s,\pi}\cdot(0,0,-1)$. Then

$$\tilde{Y}_1 = \begin{pmatrix} -x \\ \sin y \\ 1-\cos y \end{pmatrix} + r_1\begin{pmatrix} a\sinh(x\sinh c) \\ -\cosh c\ \cos y\cos(y\cosh c) - \sin y\sin(y\cosh c) \\ -\cosh c\ \sin y\cos(y\cosh c) + \cos y\sin(y\cosh c) \end{pmatrix},$$

$$\tilde{Y}_2 = \begin{pmatrix} -x \\ -\sin y \\ -1+\cos y \end{pmatrix} + r_2\begin{pmatrix} a\sinh(x\sinh c) \\ -\cosh c\ \cos y\cos(y\cosh c) - \sin y\sin(y\cosh c) \\ -\cosh c\ \sin y\cos(y\cosh c) + \cos y\sin(y\cosh c) \end{pmatrix},$$

where $a = \{(z_1^2 - w_1^2)/(w_3^2 + w_2^2\operatorname{sech}^2 c)\}^{\frac{1}{2}}$, and

$$r_1 = \frac{2}{\sinh c\{a\cosh(x\sinh c) + \sinh c\ \sin(y\cosh c)\}},$$

$$r_2 = \frac{2}{\sinh c\{a\cosh(x\sinh c) - \sinh c\ \sin(y\cosh c)\}}.$$

If $a > \sinh c$ and $\cosh c$ is a rational number n/k, then $\tilde{Y}_1, \tilde{Y}_2$ are immersed cylinder with n bubbles (see Figures 5, 6, 7).

Note that $\tilde{u}$ and the principal curvatures $\tilde{k}_{i,1}$, $\tilde{k}_{i,2}$ of $\tilde{Y}_i$ are given by

$$\begin{aligned}
e^{\tilde{u}} &= \frac{a\cosh(x\sinh c) + \sinh c\ \sin(y\cosh c)}{a\cosh(x\sinh c) - \sinh c\ \sin(y\cosh c)},\\
\tilde{k}_{1,1} &= -\frac{2a\sinh c\cosh(x\sinh c)\sin(y\cosh c)}{(a\cosh(x\sinh c) - \sinh c\ \sin(y\cosh c))^2},\\
\tilde{k}_{1,2} &= \frac{a^2\cosh(x\sinh c) + \sinh^2 c\ \sin(y\cosh c)}{(a\cosh(x\sinh c) - \sinh c\ \sin(y\cosh c))^2},\\
\tilde{k}_{2,1} &= -\frac{2a\sinh c\cosh(x\sinh c)\sin(y\cosh c)}{(a\cosh(x\sinh c) + \sinh c\ \sin(y\cosh c))^2},\\
\tilde{k}_{2,2} &= -\frac{a^2\cosh(x\sinh c) + \sinh^2 c\ \sin(y\cosh c)}{(a\cosh(x\sinh c) + \sinh c\ \sin(y\cosh c))^2}.
\end{aligned}$$

Note that $\tilde{H}_1 = 1$ and $\tilde{H}_2 = -1$. This does not contradict Corollary 13.3 because the isothermic coordinate system (x, y) is not the canonical one for the cylinder viewed as a CMC surface. The canonical isothermic coordinates are $x/2, y/2$.

14. Bäcklund transformations and loop group factorizations

Terng and Uhlenbeck ([TU2]) show that the classical Bäcklund transformation of surfaces in R^3 with constant curvature -1 corresponds to the action of a rational map with only one simple pole on the space of solutions of the SGE. In this section, we give a generalization of this result to the $G_{n,n}$-system. We find a rational map that satisfies the $G_{n,n}$-reality condition and has only one simple pole, and show that the corresponding geometric action gives rise to a Bäcklund transformation of n-dimensional submanifolds.

Bäcklund's theorem is generalized by Tenenblat and Terng in [TT] to submanifolds in R^{2n-1}, and by Tenenblat in [Ten] to submanifolds in S^{2n-1} and H^{2n-1}. These generalizations arise naturally if we reformulate the classical Bäcklund transformations in terms of orthonormal frames as follows: Note that $\ell : M \to \tilde{M}$ is a Bäcklund transformation with constant θ for surfaces $M, \tilde{M}$ in R^3, then there exist $O(3)$-frame e_A and $\tilde{e}_A$ such that $\ell(p) = p + \sin\theta\, e_1(p)$ and

$$(\tilde{e}_1, \tilde{e}_2, \tilde{e}_3)(\ell(x)) = (e_1, e_2, e_3)(x)\begin{pmatrix} 1 & 0 & 0\\ 0 & \cos\theta & -\sin\theta\\ 0 & \sin\theta & \cos\theta \end{pmatrix}$$

for all $x \in M$.

14.1 Definition ([Ten]). Let $M, \tilde{M}$ be n-dimensional submanifolds in S^{2n-1}. A diffeomorphism $\ell : M \to \tilde{M}$ is a *Bäcklund transformation* with constant θ if there exist local orthonormal $O(2n)$-frames $\{e_A\}, \{\tilde{e}_A\}$ of $M, \tilde{M}$ respectively such that

(i) $e_1 = X$ and $\tilde{e}_1 = \tilde{X}$ are the immersions,

(i) $\{e_\alpha\}_{\alpha=n+1}^{2n-1}$ and $\{\tilde{e}_\alpha\}_{\alpha=n+1}^{2n-1}$ are parallel normal frames for $M, \tilde{M}$ respectively,

(iii)

$$\begin{aligned}&(\tilde{X}, \tilde{e}_{n+1}, \cdots, \tilde{e}_{2n-1}, \tilde{e}_1, \tilde{e}_n)(\ell(x)) \\ &= (X, e_{n+1}, \cdots, e_{2n-1}, e_1, \cdots, e_n)(x)\begin{pmatrix} \cos\theta\, I_n & -\sin\theta\, I_n \\ \sin\theta\, I_n & \cos\theta\, I_n \end{pmatrix}\end{aligned}$$

for all $x \in M$.

14.2 Theorem ([Ten]). *If $\ell : M^n \to \tilde{M}^n$ is a Bäcklund transformation in S^{2n-1}, then both M and $\tilde{M}$ are flat. Moreover, if M^n is a flat submanifold of S^{2n-1}, then given any constant θ and a unit vector $v_0 \in TM_{p_0}$, there exist a submanifold $\tilde{M}$ of S^{2n-1} and a Bäcklund transformation $\ell : M \to \tilde{M}$ such that $\ell(p_0) = \cos\theta\, \ell(p_0) + \sin\theta\, v_0$.*

Next, we explain the relation between the geometric Bäcklund transformation and the dressing action on the space of $G_{n,n}$-systems. By Theorem 6.1, the Gauss-Codazzi equations for flat n-dimensional submanifold in S^{2n-1} are the $G_{n,n}$-system I, which is gauge equivalent to the $G_{n,n}$-system for $F : R^n \to gl_*(n)$ such that

$$\theta_\lambda = \lambda\begin{pmatrix} 0 & -\delta \\ \delta & 0 \end{pmatrix} + \begin{pmatrix} -F\delta + \delta F^t & 0 \\ 0 & -F^t\delta + \delta F \end{pmatrix} \tag{14.1}$$

is flat for all $\lambda \in C$, where $\delta = \mathrm{diag}(dx_1, \cdots, dx_n)$. So the $G_{n,n}$-system is the equation for F:

$$\begin{cases} C_iF^t_{x_j} - F_{x_j}C_i - C_jF^t_{x_i} + F_{x_i}C_j = [C_iF^t - FC_i, C_jF^t - FC_j], \\ C_iF_{x_j} - F^t_{x_j}C_i - C_jF_{x_i} + F^t_{x_i}C_j = [C_iF - F^tC_i, C_jF - F^tC_j], \end{cases} \tag{14.2}$$

where $C_i = e_{ii} = \mathrm{diag}(0, \cdots, 1, 0, \cdots, 0)$ as before.

Let $s \in R$ be a non-zero constant, $\beta \in O(n)$ a constant, and

$$k_{s,\beta}(\lambda) = \frac{1}{\lambda - is}\begin{pmatrix} s\beta & \lambda \\ -\lambda & s\beta^t \end{pmatrix}. \tag{14.3}$$

Note that $k_{s,\beta}$ is holomorphic at $\lambda = \infty$, $k_{s,\pi}(\infty) \neq I$, and $k_{s,\beta}$ only satisfies the $G_{n,n}$-reality condition up to a scalar function, i.e.,

$$\overline{k(\bar{\lambda})} = \frac{\lambda - is}{\lambda + is}\, k(\lambda), \quad I_{n,n}k(\lambda)I_{n,n} = k(-\lambda), \quad k(\lambda)^t k(\lambda) = \frac{\lambda^2 + s^2}{(\lambda - is)^2}\, I,$$

where $k = k_{s,\beta}$. So $k_{s,\beta}$ does not belong to G_-. But the following is true: (i) an element of the form $f(\lambda)I$ in G_- lies in the center, (ii) the dressing action of such element on the space of solutions of the $G_{n,n}$-system is trivial, and (iii) the factorization still works. In fact, if E is a frame of a solution F of the $G_{n,n}$-system (14.2), then $k_{s,\pi}(\lambda)E(x, \lambda)$ can still be factored as $\tilde{E}(x, \lambda)k_{s,\tilde{\beta}(x)}$ for some functions $\tilde{\beta}(x)$ and $\tilde{E}(x, \lambda)$ so that E is holomorphic for $\lambda \in C$. Hence we get

14.3 Theorem. *Let F be a solution of the $G_{n,n}$-system (14.2), E a frame of F, $s \in R$ a constant, $\beta \in O(n)$ a constant matrix, and $k_{s,\beta}$ defined by (14.3). Write $E(x,-is) = \begin{pmatrix} \eta_1(x) & \eta_2(x) \\ \eta_3(x) & \eta_4(x) \end{pmatrix}$ with $\eta_i \in \mathcal{M}_{n\times n}$. Set*

$$\begin{aligned} \tilde{\beta} &= (i\eta_4 - \beta\eta_2)^{-1}(i\beta\eta_1 + \eta_3), \\ \tilde{E}(x,\lambda) &= k_{s,\beta}E(x,\lambda)k_{s,\tilde{\beta}(x)}(\lambda)^{-1}, \\ \tilde{F} &= F^t + s\tilde{\beta}_*, \end{aligned} \tag{14.4}$$

where y_* is the matrix whose (i,j)-th entry is y_{ij} for $i \neq j$ and is 0 for $i = j$. Then
(i) $\tilde{F}$ is a solution of the $G_{n,n}$-system (14.2) and $\tilde{E}$ is a frame of $\tilde{F}$,
(ii) $\tilde{\beta}$ is a solution of

$$\begin{cases} d\tilde{\beta} = \sum_{i=1}^n (-\tilde{\beta}(FC_i - C_iF^t) + (F^tC_i - C_iF)\tilde{\beta} - sC_i + s\tilde{\beta}C_i\tilde{\beta})dx_i, \\ \tilde{\beta}\tilde{\beta}^t = I \end{cases}. \tag{14.5}$$

PROOF. First we prove that $\tilde{E}(x,\lambda)$ is holomorphic in $\lambda \in C$. By definition, $\tilde{E}(x,\lambda)$ is holomorphic for all $\lambda \in C$ except at $\lambda = \pm is$, and has simple poles at is and $-is$. A direct computation implies that the residue at $-is$ is a constant times

$$\begin{pmatrix} \beta & -iI_n \\ iI_n & \beta^t \end{pmatrix} \begin{pmatrix} \eta_1 & \eta_2 \\ \eta_3 & \eta_4 \end{pmatrix} \begin{pmatrix} \tilde{\beta}^t & iI_n \\ -iI_n & \tilde{\beta} \end{pmatrix},$$

which is zero by the definition of $\tilde{\beta}$. Similar computation shows that the residue of $\tilde{E}(x,\lambda)$ at $\lambda = is$ is also zero. Hence $\tilde{E}(x,\lambda)$ is holomorphic in $\lambda \in C$.

Let $\tilde{\theta}_\lambda = \tilde{E}^{-1}d\tilde{E}$, and $\theta_\lambda = E^{-1}dE$. Then $\tilde{\theta}_\lambda$ is holomorphic for $\lambda \in C$. It remains to prove that $\tilde{\theta}_\lambda$ is of the form (2.3) for some $\tilde{F}$. But

$$\tilde{\theta}_\lambda = \tilde{E}^{-1}d\tilde{E} = k_{s,\tilde{\beta}}\theta_\lambda k_{s,\tilde{\beta}}^{-1} - dk_{s,\tilde{\beta}}k_{s,\tilde{\beta}}^{-1}. \tag{14.6}$$

Because $k_{s,\tilde{\beta}}$ is holomorphic at $\lambda = \infty$ and θ_λ has only simple poles at $\lambda = \infty$, so is $\tilde{\theta}_\lambda$. Compare coefficients of λ^i in (14.6) to conclude that $\tilde{\theta}_\lambda$ is of the form (2.3) for some $\tilde{F}$ and $\tilde{F}$ is given by (14.4). This proves (i).

Multiply (14.6) by $k_{s,\tilde{\beta}}$ on the right to get

$$\tilde{\theta}_\lambda k_{s,\tilde{\beta}} = k_{s,\tilde{\beta}}\theta_\lambda - dk_{s,\tilde{\beta}}.$$

Multiply the above equation by $(\lambda - is)$ then compare coefficients of λ^i to get the differential equation for $\tilde{\beta}$ in (ii). ∎

14.4 Corollary. *Let F be a solution of the $G_{n,n}$-system (14.2), and E a frame of F. Then system (14.5) is solvable for $\tilde{\beta}$. Moreover, if $\tilde{\beta}$ is a solution of (14.5) with initial condition $\tilde{\beta}(0) = \beta$, then $\tilde{F} = F^t + s\tilde{\beta}_*$ is a solution of (14.2).*

Since solutions of the $G_{n,n}$-system I correspond to flat n-submanifolds in S^{2n-1}, Theorem 14.3 and Corollary 14.4 give a method of constructing new flat n-submanifolds in S^{2n-1} from a given one. Geometrically, this gives the geometric Bäcklund transformation constructed by Tenenblat [Ten]:

14.5 Theorem. *Let $F, E, k_{s,\beta}, \tilde{\beta}, \tilde{E}, \tilde{F}$ be as in Theorem 14.3. Write $E(x,0) = \begin{pmatrix} A(x) & 0 \\ 0 & B(x) \end{pmatrix}$. Let*

$$N(x) = E^I(x,1) = E(x,1)\begin{pmatrix} A(x)^t & 0 \\ 0 & I_n \end{pmatrix},$$
$$\tilde{N}(x) = \tilde{E}^{\natural\, I}(x,1) = N(x)\begin{pmatrix} A & 0 \\ 0 & I_n \end{pmatrix}\begin{pmatrix} \cos\rho\,\tilde{\beta}^t & -\sin\rho\, I_n \\ \sin\rho\, I_n & \cos\rho\,\tilde{\beta} \end{pmatrix}\begin{pmatrix} \tilde{A}^{\natural\, t} & 0 \\ 0 & I_n \end{pmatrix}, \tag{14.7}$$

where $\rho = \arctan(1/s)$, and $\tilde{A}^\natural = A\tilde{\beta}^t$. Let v_i and $\tilde{v}_i$ denote the i-th column of N and $\tilde{N}$ respectively. Then the map $v_1(x) \mapsto \tilde{v}_1(x)$ is a Bäcklund transformation of n-dimensional submanifolds in S^{2n-1} with constant ρ defined by Definition 14.1.

PROOF. By (14.4),

$$\tilde{E}(x,0) = \begin{pmatrix} \beta & 0 \\ 0 & \beta^t \end{pmatrix}\begin{pmatrix} A(x)\tilde{\beta}^t(x) & 0 \\ 0 & B(x)\tilde{\beta}(x) \end{pmatrix}.$$

Since $\begin{pmatrix} \beta & 0 \\ 0 & \beta^t \end{pmatrix}$ is constant, $\begin{pmatrix} A\tilde{\beta}^t & 0 \\ 0 & B\tilde{\beta} \end{pmatrix}$ is also a trivialization of the Lax connection (14.1) for $\tilde{F}$ at $\lambda = 0$. By Proposition 3.6, (A, F) and $(\tilde{A}, \tilde{F})$ are solutions of the U/K-system I and $N, \tilde{N}$ are the corresponding frames at $\lambda = 1$. It follows from Theorem 6.1 that $v_1, \tilde{v}_1$ are flat n-dimensional submanifolds of S^{2n-1}, $(v_2, \cdots, v_n)$ and $(\tilde{v}_2, \cdots, \tilde{v}_n)$ are parallel normal frames for v and $\tilde{v}_1$ respectively. Multiply both sides of (14.7) by $\begin{pmatrix} I_n & 0 \\ 0 & A^t \end{pmatrix}$ on the right to get

$$\begin{aligned} \tilde{N}\begin{pmatrix} I_n & 0 \\ 0 & A^t \end{pmatrix} &= N\begin{pmatrix} \cos\rho I_n & -\sin\rho I_n \\ \sin\rho\,\tilde{\beta}A^t & \cos\rho\tilde{\beta}A^t \end{pmatrix} \\ &= N\begin{pmatrix} I_n & 0 \\ 0 & \tilde{\beta}A^t \end{pmatrix}\begin{pmatrix} \cos\rho\, I_n & -\sin\rho\, I_n \\ \sin\rho\, I_n & \cos\rho\, I_n \end{pmatrix}. \end{aligned} \tag{14.8}$$

Let

$$\tilde{N}_b = \tilde{N}\begin{pmatrix} I_n & 0 \\ 0 & A^t \end{pmatrix}, \quad N_b = N\begin{pmatrix} I_n & 0 \\ 0 & \tilde{A}^{\natural\, t} \end{pmatrix}.$$

Then (14.8) can be rewritten as

$$\tilde{N}_b = N_b\begin{pmatrix} \cos\rho\, I_n & -\sin\rho\, I_n \\ \sin\rho\, I_n & \cos\rho\, I_n \end{pmatrix}. \tag{14.9}$$

The first n column vectors of N and N_b are the same, the first n column vectors of $\tilde{N}$ and $\tilde{N}_b$ are the same and they are parallel normal frames. The last n columns of N_b and $\tilde{N}_b$ are tangent frames for v_1 and $\tilde{v}_1$ respectively (they are not principal curvature directions). Geometrically, (14.9) means that the map $v_1 \mapsto \tilde{v}_1$ is a Bäcklund transformation with constant ρ. ∎

14.6 Example. We apply Theorem 14.5 to the trivial solution $F = 0$ to get explicit immersions of flat n-submanifolds in S^{2n-1}. The Lax connection of $F = 0$ of the $G_{n,n}$-system (14.1) is

$$\theta_\lambda = \lambda \sum_i \begin{pmatrix} 0 & -C_i \\ C_i & 0 \end{pmatrix} dx_i.$$

So

$$E(x, \lambda) = \begin{pmatrix} C(x, \lambda) & -S(x, \lambda) \\ S(x, \lambda) & C(x, \lambda) \end{pmatrix}$$

is a frame of $F = 0$, where $C(x, \lambda) = \text{diag}(\cos(\lambda x_1), \ldots, \cos(\lambda x_n))$ and $S(x, \lambda) = \text{diag}(\sin(\lambda x_1), \cdots, \sin(\lambda x_n))$. Note that

$$E(x, 0) = \begin{pmatrix} I_n & 0 \\ 0 & I_n \end{pmatrix} = \begin{pmatrix} A & 0 \\ 0 & B \end{pmatrix},$$

so $A = I_n$. It follows from Theorem 14.3 that

$$\begin{aligned} &\tilde{\beta} = (p_1 - \beta p_2)^{-1}(\beta p_1 - p_2), \quad \text{where} \\ &p_1 = \text{diag}(\cosh(sx_1), \cdots, \cosh(sx_n)), \quad p_2 = \text{diag}(\sinh(sx_1), \cdots, \sinh(sx_n)). \end{aligned}$$

Since $A = I_n$, $\tilde{A}^\natural = A\tilde{\beta}^t = \tilde{\beta}^t$ and formula (14.7) give

$$\tilde{N}(x) = \begin{pmatrix} C(x, 1) & -S(x, 1) \\ S(x, 1) & C(x, 1) \end{pmatrix} \begin{pmatrix} \cos \rho\, I_n & -\sin \rho\, I_n \\ \sin \rho\, \tilde{\beta} & \cos \rho\, \tilde{\beta} \end{pmatrix}.$$

The first column of $\tilde{N}(x)$ gives an explicit immersion of flat n-submanifolds in S^{2n-1} with flat and non-degenerate normal bundle.

Next we recall a generalization of Bäcklund's theorem to n-dimensional submanifolds in R^{2n-1} with constant sectional curvature -1 proved by Tenenblat and Terng in [TT]. First we recall the definition.

14.7 Definition ([TT]). Let $M, \tilde{M}$ be n-dimensional submanifolds in R^{2n-1} with flat normal bundle. A diffeomorphism $\ell : M \to \tilde{M}$ is called a *Bäcklund transformation* with constant θ if there exist local orthonormal frames $\{e_A\}$ and $\{\tilde{e}_A\}$ of $M, \tilde{M}$ respectively such that

(i) $\{e_\alpha\}_{\alpha=n+1}^{2n-1}$ and $\{\tilde{e}_\alpha\}_{\alpha=n+1}^{2n-1}$ are parallel normal frames,

(ii) $\ell(x) = x + \sin\theta\, e_1(x)$ for all $x \in M$,

(iii)

$$(\tilde{e}_1, \cdots, \tilde{e}_{2n-1})(\ell(x)) = (e_1, \cdots, e_{2n-1})(x) \begin{pmatrix} 1 & 0 & 0 \\ 0 & \cos\theta\, I_{n-1} & -\sin\theta\, I_{n-1} \\ 0 & \sin\theta\, I_{n-1} & \cos\theta\, I_{n-1} \end{pmatrix}$$

for all $x \in M$.

14.8 Theorem ([TT]). *If $\ell : M^n \to \tilde{M}^n$ is a Bäcklund transformation in R^{2n-1} with constant θ, then both M and $\tilde{M}$ have constant sectional curvature -1. Moreover, if M^n is a submanifold in R^{2n-1} with sectional curvature -1, then given any constant θ and a unit vector $v_0 \in TM_{p_0}$, there exist a submanifold $\tilde{M}$ of R^{2n-1} and a Bäcklund transformation $\ell : M \to \tilde{M}$ such that $\ell(p_0) = p_0 + \sin\theta\, v_0$.*

To explain the analytic version of the above theorem, we first recall that ([TT]) if M is an n-dimensional submanifold of R^{2n-1} with sectional curvature -1 then the normal bundle is flat and there exist line of curvature coordinates such that the two fundamental forms are

$$\begin{aligned} I &= \sum_{i=1}^{n} a_{1i}^2 dx_i^2, \\ II &= \sum_{i,j=1}^{i=n,j=n-1} a_{1i}a_{ji}dx_i^2 e_{n+j-1} \end{aligned} \tag{14.10}$$

for some $O(n)$-valued map $A = (a_{ij})$ and parallel orthonormal normal frame $e_{n+1}, \cdots, e_{2n-1}$. The Gauss-Codazzi equations for M are the following system for (A, F):

$$\begin{cases} A^{-1}dA = -F\delta + \delta F^t, \\ d\omega + \omega \wedge \omega = -\delta A^t e_{11} A\delta, \quad \text{where} \end{cases}$$
$$\delta = \text{diag}(dx_1, \cdots, dx_n), \ \omega = \delta F - F^t\delta, \ \text{and} \ \ e_{11} = \text{diag}(1, 0, \cdots, 0). \tag{14.11}$$

Equation (14.11) is a called the *generalized sine-Gordon equation* (GSGE). The analytic version of Theorem 14.8 is

14.9 Theorem ([TT]). *Let (A, F) be a solution of the GSGE (14.11). Then the following equation for $\tilde{A}$ is solvable:*

$$d\tilde{A} = -\tilde{A}(-F^t\delta + \delta F) + \tilde{A}\delta A^t D\tilde{A} - DA\delta. \tag{14.12}$$

Moreover,
(i) for all $j \neq k$ and i, we have $(\tilde{a}_{ij})_{x_k}/\tilde{a}_{ik} = (\tilde{a}_{rj})_{x_k}/\tilde{a}_{rk}$, which will be denoted by $\tilde{f}_{jk}$,
(ii) let $\tilde{F}$ be the $gl(n)$-valued map whose jk-th entry is $\tilde{f}_{jk}$ for $j \neq k$ and $\tilde{f}_{jj} = 0$, then $(\tilde{A}, \tilde{F})$ is again a solution of the GSGE (14.11).

Next we explain the relation between GSGE and the dressing action of $G^1_{n,n}$-systems. It follows from Theorem 7.3 that local isometric immersions of H^n in R^{2n-1} correspond to solutions (A, F, b^t) of the $G^1_{n,n}$-system I such that $b^t = Ae_{11}$. It can be easily seen that the GSGE (14.11) is the $G^1_{n,n}$-system (4.2) with $v = (F, b^t)$. Hence GSGE has a Lax connection

$$\theta_\lambda = \begin{pmatrix} \omega & -\delta\lambda & 0 \\ \delta\lambda & \tau & \delta b^t \\ 0 & b\delta & 0 \end{pmatrix}, \quad \text{where } \omega = -F\delta + \delta F^t, \ \tau = -F^t\delta + \delta F. \tag{14.13}$$

However, unlike the $G_{n,n}$-system, we are not able to find a rational map with only one simple pole satisfying the $G^1_{n,n}$-reality condition up to scalar function. But the following element satisfies the first two conditions of the $G^1_{n,n}$-reality condition (11.1):

$$g = \lambda \begin{pmatrix} 0 & -I & 0 \\ I & 0 & 0 \\ 0 & 0 & 0 \end{pmatrix} + \begin{pmatrix} \beta & 0 & 0 \\ 0 & \gamma & \xi^t \\ 0 & \eta & \epsilon \end{pmatrix}, \tag{14.14}$$

where $\beta, \gamma, \xi, \eta, \epsilon$ are matrix valued functions on R^n. We explain below how to use the gauge transformation of g (all entries are assumed to be functions of x) of the Lax connection θ_λ of the $G^1_{n,n}$-system so that $g * \theta_\lambda = \tilde{\theta}_\lambda$ for some other solution $(\tilde{F}, \tilde{b}^t)$. Suppose there exists a solution $\tilde{v} = (\tilde{F}, \tilde{b}^t)$ of the $G^1_{n,n}$-system (14.13) such that

$$g\theta_\lambda - dg = \tilde{\theta}_\lambda g, \tag{14.15}$$

where $\tilde{\theta}$ is the Lax connection (14.13) corresponding to $\tilde{v}$. Equate coefficient of λ^j of (14.15) for each j to get

$$\begin{cases} \tilde{w} = \tau + \beta\delta - \delta\gamma, \\ \tilde{\tau} = \omega + \gamma\delta - \delta\beta, \\ \eta = \tilde{b}, \quad \xi = b, \\ d\beta = \beta\omega - \tilde{w}\beta, \\ d\gamma = \gamma\tau + \xi^t b\delta - \tilde{\tau}\gamma + \delta\tilde{b}^t\eta, \\ d\xi^t = \gamma\delta b^t - \tilde{\tau}\xi^t - \delta\tilde{b}^t\epsilon, \\ d\eta = \eta\tau + \epsilon b\delta - \tilde{b}\delta\gamma, \\ d\epsilon = 0. \end{cases} \tag{14.16}$$

The first and second equations of this system imply that $\beta = \gamma^t$, and the fourth and fifth equations imply that

$$\delta(\beta^t\beta - b^t b) = (\beta\beta^t - \tilde{b}^t\tilde{b})\delta.$$

Equate each entry of the above equation to get

$$\beta^t\beta - b^t b = \beta\beta^t - \tilde{b}^t\tilde{b} = \triangle \tag{14.17}$$

for some diagonal matrix. We make an ansatz that $\triangle = rI$ for some constant r. Note that the fourth equation of (14.16) means that $\tilde{w}$ is the gauge transformation of ω by β. Hence there exists a constant matrix C so that the trivialization A of ω and $\tilde{A}$ of $\tilde{w}$ are related by

$$CA = \tilde{A}\beta. \tag{14.18}$$

Substitute (14.18) into (14.17) to get

$$A^t(C^tC - e_{11})A = \tilde{A}^t(CC^t - e_{11})\tilde{A} = rI.$$

Hence $C^tC = CC^t = rI + e_{11}$. Let $r = \cot^2\theta$. Then $\beta = \tilde{A}^t DA$, where $D = \mathrm{diag}(\csc\theta, \cot\theta, \cdots, \cot\theta)$. Substitute this to the fourth equation of (14.16), we get the analytic Bäcklund transformation (14.12).

Notice that the element g defined as in (14.14) can be written as

$$g = \begin{pmatrix} \tilde{A}^t & 0 & 0 \\ 0 & A^t & 0 \\ 0 & 0 & 1 \end{pmatrix} \begin{pmatrix} D & -\lambda I & 0 \\ \lambda I & D & e_1^t \\ 0 & e_1 & \csc\theta \end{pmatrix} \begin{pmatrix} A & 0 & 0 \\ 0 & \tilde{A} & 0 \\ 0 & 0 & 1 \end{pmatrix},$$

where $e_1 = (1, 0, \cdots, 0)$ and $D = \operatorname{diag}(\csc\theta, \cot\theta, \cdots, \cot\theta)$.

15. Permutability Formula for Ribaucour transformations

Bianchi proved a Permutability Theorem for Bäcklund transformations for surfaces in R^3: If $\ell : M_0 \to M_i$ is a Bäcklund transformation with constant θ_i for surfaces in R^3 for $i = 1, 2$ and $\sin^2\theta_1 \neq \sin^2\theta_2$, then there exist a unique surface M_3 and Bäcklund transformations $\tilde{\ell}_1 : M_2 \to M_3$ and $\tilde{\ell}_2 : M_1 \to M_3$ with constant θ_1, θ_2 respectively such that $\tilde{\ell}_1 \circ \ell_2 = \tilde{\ell}_2 \circ \ell_1$. Moreover, if q_i is the solution of the SGE corresponding to M_i respectively, then

$$\tan\frac{q_3 - q_0}{4} = \frac{s_1 + s_2}{s_1 - s_2} \tan\frac{q_1 - q_2}{4},$$

where $s_i = \tan(\theta_i/2)$. Terng and Uhlenbeck proved in [TU2] that the Permutability theorem is a consequence of a relation between generators $h_{s,\pi}$ defined by (9.4) and fact that the dressing is a group action. In this section, we find the relation among $g_{s,\pi}$'s and use the same proof as in [TU2] to get the Permutability Theorem for Ribaucour transformations for submanifolds associated to $G_{m,n}$- and $G^1_{m,n}$-systems.

Recall that given unit vectors W in R^m and Z in R^n, $g_{s,\pi}$ was defined by

$$\begin{aligned} g_{s,\pi}(\lambda) &= \left(\pi + \frac{\lambda - is}{\lambda + is}(I - \pi)\right)\left(\bar{\pi} + \frac{\lambda + is}{\lambda - is}(I - \bar{\pi})\right) \\ &= I + \frac{2is\pi}{\lambda - is} - \frac{2is\bar{\pi}}{\lambda + is}. \end{aligned}$$

where π is the Hermitian projection of C^{n+m} onto $\mathbf{C}\begin{pmatrix} W \\ iZ \end{pmatrix}$.

15.1 Proposition. *Let W_k and Z_k be unit vectors in R^m, R^n respectively, $v_k^t = (W_k^t, iZ_k^t)$, and π_k the Hermitian projections onto v_k for $k = 1, 2$ respectively. Let $s_1, s_2 \in R$ be constants such that $s_1^2 \neq s_2^2$ and $s_1 s_2 \neq 0$. Let u_k denote the unit direction of*

$$g_{s_j,\pi_j}(-is_k)(v_k)$$

for $j \neq k$, and τ_k the Hermitian projection onto u_k. Then:

(i) u_k is of the form $\frac{1}{\sqrt{2}}(U_k^t, iV_k^t)$ for some unit vectors $U_k \in R^n$ and $V_k \in R^m$.

(ii)

$$g_{s_2,\tau_2}(\lambda) g_{s_1,\pi_1}(\lambda) = g_{s_1,\tau_1}(\lambda) g_{s_2,\pi_2}(\lambda). \tag{15.1}$$

(iii) τ_1, τ_2 are unique projections satisfying (15.1).

PROOF.

(i) follows from the fact that g_{s_j,π_j} satisfies the $G_{m,n}$-reality condition.

(ii) Let $g_i = g_{s_i,\pi}$ and $\tilde{g}_i = g_{s_i,\tau_i}$ for $i = 1, 2$. The residue of $\tilde{g}_1 g_2 g_1^{-1}$ at $\lambda = is_1$ is

$$R_{is_1} = 2is_1(\tau_1 g_2(is_1)(I - \pi_1) + (I - \bar{\tau}_1)g_2(is_1)\bar{\pi}_1).$$

Since u_1 is parallel to $g_2(-is_1)(v_1)$,

$$\langle g_2(is_1)(v_1^\perp), u_1\rangle = \langle v_1^\perp, g_2(is_1)^* g_2(-is_1)(v_1)\rangle,$$

where $v_1^\perp$ is the orthogonal complement of v_1. But $g(\bar{z})^* g(z) = I$. So the above inner product is zero, i.e., $\tau_1 g_2(is_1)(I - \pi_1) = 0$. Since $\bar{u}_1$ is parallel to

$$\overline{g_2(-is_1)(v_1)} = g_2(is_1)(\bar{v}_1),$$

a similar argument gives $(I - \bar{\tau}_1)g_2(is_1)\bar{\pi}_1 = 0$. Hence $R_{is_1} = 0$, which implies that $\tilde{g}_1 g_2 g_1^{-1}$ is holomorphic at $\lambda = is$. Since $\tilde{g}_1 g_2 g_1^{-1}$ satisfies the reality condition, it is also holomorphic at $\lambda = -is_1$. So $h_1 = \tilde{g}_1 g_2 g_1^{-1} \tilde{g}_2^{-1}$ is holomorphic at $\pm is_1$. Use a similar argument to prove that $h_2 = \tilde{g}_2 g_1 g_2^{-1} \tilde{g}_1^{-1}$ is holomorphic at $\lambda = \pm is_2$. But $h_1 = h_2^{-1}$, which is holomorphic for all $\lambda \in C$ except at $\lambda = \pm is_1, \pm is_2$. Since $h_i(\lambda)^{-1} = h_i(\bar{\lambda})^*$, h_2^{-1} is holomorpic at $\pm is_2$. This proves that h_1 is holomorphic for all $\lambda \in C$. But $h_1(\infty) = I$. So $h_1 = I$. This proves (ii).

(iii) Let σ_i denote the projection onto y_i, and $\phi_i = g_{s_i,\sigma_i}$. Suppose $\phi_1 g_2 = \phi_2 g_1$. We want to prove $\sigma_i = \tau_i$. But $\phi_1 = \phi_2 g_1 g_2^{-1}$ implies that $\phi_2 g_1 g_2^{-1}$ is holomorphic at $\lambda = is_2$. Hence the residue at is_2 must be zero, i.e.,

$$\sigma_2 g_1(is_2)(I - \pi_2) + (I - \bar{\sigma}_2)g_2(is_2)\bar{\pi}_2 = 0. \tag{15.2}$$

Since

$$\langle v_2, \bar{v}_2\rangle = 0, \quad \langle y_2, \bar{y}_2\rangle = 0,$$

(15.2) implies that $\sigma_2 g_1(is_2)(I - \pi_2) = 0$. So

$$0 = \langle g_1(is_2)(v_2^\perp), y_2\rangle = \langle v_2^\perp, g_1(is_2)^*(y_2)\rangle = \langle v_2^\perp, g_1(-is_2)^{-1}(y_2)\rangle.$$

This proves that $g_1(-is_2)^{-1}(y_2) \in Cv_2$. Hence y_2 is parallel to $g_1(-is_2)(v_2)$. ∎

It is known (cf. [TU2]) that an analogue of Bianchi's Permutability Theorem can be obtained easily from the above Proposition. We sketch the reason here. Let ξ be a solution of the $G_{m,n}$-system, and E the frame of ξ with $E(0, \lambda) = I$. Let $g_j = g_{is_j,\pi_j}$, $h_j = g_{is_j,\tau_j}$ as in Proposition 15.1. So $h_1 g_2 = h_2 g_1$. Since dressing $\sharp$ is an action,

$$(h_1 g_2)\sharp\xi = (h_2 g_1)\sharp\xi.$$

Let $\xi_i = g_i \sharp \xi$. Then

$$h_1 \sharp \xi_2 = h_2 \sharp \xi_1,$$

which will be denoted by ξ_3. Factor

$$g_1 E = E^1 \tilde{g}_1, \quad g_2 E = E^2 \tilde{g}_2, \quad h_2 E^1 = E^4 \tilde{h}_2, \quad h_1 E^2 = E^3 \tilde{h}_1$$

such that $E^i(x,\lambda)$ are holomorphic for $\lambda \in C$ and $\tilde{g}_i(x,\lambda), \tilde{h}_i(x,\lambda)$ are meromorphic for $\lambda \in S^2$ and are equal to I at $\lambda = \infty$. Then

$$h_2 g_1 E = E^4 \tilde{h}_2 \tilde{g}_1, \quad h_1 g_2 E = E^3 \tilde{h}_1 \tilde{g}_2.$$

It follows from the assumption $h_1 g_2 = h_2 g_1$ and the uniqueness of factorization that $E^3 = E^4$, and $\tilde{h}_1 \tilde{g}_2 = \tilde{h}_2 \tilde{g}_1$. So by Proposition 15.1 and Theorem (7.10), we have

$$g_{s_1,\tilde{\tau}_1(x)} g_{s_2,\tilde{\pi}_2(x)} = g_{s_2,\tilde{\tau}_2(x)} g_{s_1,\tilde{\pi}_1(x)},$$

and the projections $\tilde{\tau}_i$ and $\tilde{\pi}_i(x)$ are related the same way as τ_i and π_i.

To summarize, let ξ be a solution of the $G_{m,n}$-system, and E the frame of ξ such that $E(0,\lambda) = I$. Let π_k be the projection onto v_k, $\xi_i = g_{s_i,\pi} \sharp \xi$, and E_i frame of ξ_i.

$$\tilde{v}_k(x) = E(x, -is_k)^{-1}(v_k), \quad \hat{v}_k = \tilde{v}_k / \parallel \tilde{v}_k \parallel,$$

and $\tilde{\pi}_k(x)$ the projection onto $\hat{v}_k$. Let $\Phi : gl(n+m, C) \to \mathcal{M}_{m\times n}$ be the map defined by

$$\Phi_{m,n}\left(\begin{pmatrix} P & Q \\ R & S \end{pmatrix}\right) = Q,$$

where a $gl(n+m)$ matrix is blocked into (m,n) blocks. Then

$$\xi_k = \xi - 4is\Phi_{m,n}(\hat{v}_k \hat{v}_k^*), \quad k = 1, 2.$$

Suppose $s_1^2 \neq s_2^2$. Let

$$\tilde{u}_k = g_{is_j,\tilde{\pi}_j(x)}(-is_k)(\hat{v}_k), \quad k \neq j.$$

Then we get the Permutability Formula:

$$\begin{aligned} \xi_3 &= \xi - 4i\Phi_{m,n}(s_1 \hat{u}_1 \hat{u}_1^* + s_2 \hat{v}_2 \hat{v}_2^*) \\ &= \xi - 4i\Phi_{m,n}(s_2 \hat{u}_2 \hat{u}_2^* + s_1 \hat{v}_1 \hat{v}_1^*) \\ &= g_{is_2,\tau_2} \sharp \xi_1 = g_{is_1,\tau_1} \sharp \xi_2, \end{aligned}$$

where τ_j is the projection onto $\hat{u}_j(0)$. We have shown in section 10 that the action $g_{s,\pi}$ on the space of solutions of the $G_{m,n}$-system gives rise to Ribaucour transformations for submanifolds. Hence we get

15.2 Corollary. *Let $P_i : \nu(M) \to \nu(M_i)$ be the Ribaucour transformation for flat n-submanifolds in S^{m+n-1} corresponding to the action of g_{s_i,π_i}. If $s_1 s_2 \neq 0$ and $s_1^2 \neq s_2^2$, then there exist unique flat n-submanifold M_3 in S^{n+m-1} and Ribaucour transformations $\tilde{P}_1 : \nu(M_2) \to \nu(M_3)$ and $\tilde{P}_2 : \nu(M_1) \to \nu(M_3)$ such that $\tilde{P}_1 \circ P_2 = \tilde{P}_2 \circ P_1$.*

The Permutability Formula and Theorem for flat n-submanifolds in R^{m+n}, n-tuples in R^m of type $O(n)$, $G^1_{m,n}$-systems, and the submanifolds associated to the $G^1_{m,n}$-system can be obtained exactly by the same way. So we will not write them down.

16. The U/K-hierarchy and Finite type solutions

In this section, we give a short review how the U/K-system fits into the U/K-hierarchy of soliton equations. We also explain the relation between the ODE method of constructing finite type solutions of the U/K-system and the dressing action.

First we give a quick review of the ZS-AKNS $n \times n$ hierarchy of commuting flows. Let a_1 be a diagonal matrix with distinct eigenvalues, and $a_1, \cdots, a_{n-1}$ linearly independent diagonal matrices in $sl(n, C)$. Since a_1 has distinct eigenvalues, the map $\mathrm{ad}(a_1)(x) = [a_1, x]$ is a linear isomorphism of $sl_*(n, C)$, where $sl_*(n, C)$ is the space of all $x \in sl(n, C)$ with zero on all diagonal entries. It is known (cf. [Sa], [TU1]) that given $1 \le i \le n-1$ and a positive integer j there exists a polynomial differential operator $Q_{a_i,j}$ of order $j-1$ for maps $u : R \to sl_*(n, C)$ such that

(i) $Q_{a_i,0}(u) = a_i$ and $Q_{a_i,1}(u) = \mathrm{ad}(a_i)\,\mathrm{ad}(a_1)^{-1}(u)$,

(ii) $(Q_{a_i,j}(u))_x + [u, Q_{a_i,j}(u)] = [Q_{a_i,j+1}(u), a]$ for all j.

The j-th flow on $C^\infty(R, sl_*(n, C))$ defined by a_j is the following evolution equation

$$u_t = (Q_{a_i,j}(u))_x + [u, Q_{a_i,j}(u)]. \tag{16.1}$$

All these flows commute. The hierarchy of these commuting flows is called the *$SL(n, C)$-hierarchy*. It follows from (ii) that u is a solution of (16.1) if and only if

$$\Omega_\lambda = (a_1\lambda + u)dx + \left(\sum_{k=0}^{j} Q_{a_i,k}\lambda^{j-k}\right) dt$$

is flat for all $\lambda \in C$.

For example, when $j = 1$, equation (16.1) is

$$u_t = \mathrm{ad}(a_i)\,\mathrm{ad}(a_1)^{-1}(u_x) + [u, \mathrm{ad}(a_i)\,\mathrm{ad}(a_1)^{-1}(u)]. \tag{16.2}$$

Many well-known soliton hierarchies come from restricting the flows in the $SL(n, C)$-hierarchy to various invariant submanifolds. If $a_i \in su(n)$, then $\mathcal{M}_1 = C^\infty(R, su_*(n))$ is invariant under the j-th flow for all $j \ge 1$, where $su_*(n) = sl_*(n, C) \cap su(n)$. The restriction of the $SL(n, C)$-hierarchy of flows to $\mathcal{M}_1$ is called the $SU(n)$-hierarchy. For example, the second flow in the $SU(2)$-hierarchy is the non-linear Schrödinger equation:

$$q_t = \frac{i}{2}\left(q_{xx} + 2\,|\,q\,|^2 q\right), \tag{NLS}$$

where $u = \begin{pmatrix} 0 & q \\ -\bar{q} & 0 \end{pmatrix}$ and $a_1 = \mathrm{diag}(i, -i)$.

Let $u : R^2 \to su_*(n)$ be a solution of the j-th flow of the $SU(n)$-hierarchy, and E the solution for $E^{-1}dE = \Omega_\lambda$ with $E(0, \lambda) = I$. Then Ω_λ and E satisfy the $SU(n)$-reality condition.

If $a_i \in su(n)$ is pure imaginary for all i, then $\mathcal{M}_2 = C^\infty(R, so(n))$ is invariant under all the odd flows. The restriction of the $SU(n)$-hierarchy to $\mathcal{M}_2$ is the $SU(n)/SO(n)$-hierarchy. For example, the third flow in the $SU(2)/SO(2)$-hierarchy is the modified KdV equation:

$$q_t = -\frac{1}{4}\left(q_{xxx} + 6q^2 q_x\right),$$

where $u = \begin{pmatrix} 0 & q \\ -q & 0 \end{pmatrix}$ with q real and $a_1 = \mathrm{diag}(i, -i)$.

If $v : R^2 \to so(n)$ is a solution of the j-th flow in the $SU(n)/SO(n)$-hierarchy and E is the solution for $E^{-1}dE = \Omega_\lambda$ with $E(0, \lambda) = I$, then Ω_λ and E satisfy the $SU(n)/SO(n)$-reality condition.

We can replace $sl(n, C)$ by any complex semi-simple Lie algebra $\mathcal{G}$, $su(n)$ by a real form $\mathcal{U}$ of $\mathcal{G}$, and $SU(n)/SO(n)$ by a symmetric space U/K to construct G-, U- and U/K-hierarchies similarly.

Let G_+ denote the group of holomorphic maps from C to U_C that satisfies the U/K-reality condition (3.5), and G_- the group of germs of holomorphic map g from $\in S^2$ to U_C at $\lambda = \infty$ that satisfies the U/K-reality condition and $g(\infty) = I$ as before. Given $g \in G_-$, factor

$$g^{-1}(\lambda) \exp(a_1 \lambda x + a_i \lambda^j t) = E(x, t, \lambda) m(x, t, \lambda)^{-1}$$

with $E(x, t, \cdot) \in G_+$ and $m(x, t, \cdot) \in G_-$. It is known (cf. [TU1]) that
(i) $E^{-1}E_x = a_1\lambda + u(x, t)$ and u is a solution of the j-th flow (16.1) defined by a_i,
(ii) $Q_{a_i,k}(u)$ is the coefficient of λ^{-k} in $m^{-1}a_i m$.

Let $e_{a_i,j}(x)$ denote the one-parameter subgroup of G_+ generated by $\eta(\lambda) = a_i\lambda^j$, i.e., $e_{a_i,j}(x)(\lambda) = \exp(a_i\lambda^j x)$. Then the j-th flow defined by a_i can be viewed as the flows corresponding to the dressing action of the two parameter abelian subgroup $\{e_{a_1,1}(x)e_{a_i,j}(t) \mid (x, t) \in R^2\}$ of G_+ on G_-. The U/K-system (1.1) can be viewed as given by the n commuting first flows in the U/K- hierarchy (c.f. [TU1]). In other words, it corresponds to the dressing action of the n-dimensional abelian subgroup

$$\{e_{a_1,1}(x_1) \cdots e_{a_n,1}(x_n) \mid (x_1, \cdots, x_n) \in R^n\}$$

of G_+ on G_-. We explain this more precisely below. Let $g \in G_-$, and

$$E_0(x, \lambda) = \exp\left(\sum_{i=1}^n a_i \lambda x_i\right).$$

The Birkhoff factorization Theorem implies that there exists unique $E(x, \lambda)$ and $m(x, \lambda)$ such that

$$g(\lambda)^{-1} E_0(x, \lambda) = E(x, \lambda) m(x, \lambda)^{-1} \tag{16.3}$$

with $E(x, \cdot) \in G_+$ and $m(x, \cdot) \in G_-$ for each x. Equation (16.3) implies that $gE = E_0 m$. So we have

$$E^{-1}E_{x_i} = m^{-1}a_i m\lambda + m^{-1}m_{x_i} \tag{16.4}$$

for each $1 \leq i \leq n$. Since the left hand side of (16.4) is holomorphic for $\lambda \in C$, so is the right hand side. Expand $m(x, \lambda)$ at $\lambda = \infty$:

$$m(x, \lambda) = I + m_1(x)\lambda^{-1} + m_2(x)\lambda^{-2} + \cdots.$$

Then we have

$$m^{-1}am = a + [a, m_1]\lambda^{-1} + \cdots.$$

For $h(\lambda) = \sum_{j \leq n_0} h_j \lambda^j$, define

$$h_+(\lambda) = \sum_{j \geq 0} h_j \lambda^j, \quad h_-(\lambda) = \sum_{j \leq -1} h_j \lambda^j.$$

Since $E^{-1}E_{x_i}$ is holomorphic for $\lambda \in C$ and $m^{-1}m_{x_i} = \sum_{j \leq -1} b_j \lambda^j$, equation (16.4) implies that

$$E^{-1}E_{x_i} = (m^{-1}a_i m\lambda)_+ = a_i\lambda + [a_i, m_1], \tag{16.5}$$

$$m^{-1}m_{x_i} = -(m^{-1}a_i m\lambda)_-. \tag{16.6}$$

By Proposition 2.2, $p(m_1)$ is a solution of the U/K-system (1.1), where p is the projection onto $\mathcal{P} \cap \mathcal{A}^\perp$. In fact, $p(m_1) = g\sharp 0$, the action of g at the solution $v = 0$.

Let $\mathcal{F}$ denote the space $g \in G_-$ such that $g(\lambda)^{-1}ag(\lambda)$ is a polynomial in λ^{-1}. Solution $g\sharp 0$ is called a *finite type solution*. This is motivated by the finite type solutions constructed for CMC tori in R^3 by Pinkall and Sterling [PiS], in $N^3(c)$ by Bobenko [Bo1], and for harmonic maps from a torus to a symmetric space by Burstall, Ferus, Pedit and Pinkall [BFPP]. We explain below the relation between the dressing action and the ODE construction of finite type solutions.

First we derive a system of equations for $m^{-1}a_1 m$. It follows from a direct computation and (16.6) that

$$(m^{-1}a_1 m)_{x_i} = [m^{-1}a_1 m, m^{-1}m_{x_i}] = [m^{-1}a_1 m, -(m^{-1}a_i m\lambda)_-].$$

But $[a_1, a_i] = 0$ implies that $[m^{-1}a_1 m, m^{-1}a_i m] = 0$. So we have

$$(m^{-1}a_1 m)_{x_i} = [m^{-1}a_1 m, (m^{-1}a_i m\lambda)_+]. \tag{16.7}$$

Write

$$m^{-1}(x, \lambda)a_1 m(x, \lambda) = \sum_{j=0}^{\infty} \xi_j(x)\lambda^{-j}.$$

Then $\xi_1 = [a, m_1]$ and

$$(m^{-1}a_i m\lambda)_+ = a_i\lambda + [a_i, m_1] = a_i\lambda + \operatorname{ad}(a_i)\operatorname{ad}(a_1)^{-1}(m_1).$$

So equation (16.7) becomes

$$\left(\sum_{j=0}^{\infty} \xi_j \lambda^{-j}\right)_{x_i} = \left[\sum_{j=0}^{\infty} \xi_j \lambda^{-j}, \; a_i\lambda + \operatorname{ad}(a_i)\operatorname{ad}(a_1)^{-1}(\xi_1)\right]. \tag{16.8}$$

Since m and $a_1\lambda$ satisfy the U/K-reality condition, $m^{-1}a_1\lambda m$ satisfies the U/K-reality condition. It follows from Proposition 3.2 that the coefficient of λ^{-j} of $m^{-1}a_1 m$ satisfies the condition that $\xi_j \in \mathcal{K}$ for j odd and $\xi_j \in \mathcal{P}$ for j even.

Compare coefficient of λ^{-j} in (16.8) for each j to get

$$(\xi_j)_{x_i} = [\xi_j, \ \ \mathrm{ad}(a_i)\,\mathrm{ad}(a_1)^{-1}(\xi_1)] + [\xi_{j+1}, a_i], \qquad 0 \le j, \tag{16.9}$$

where $\xi_j \in \mathcal{K}$ if j is odd, $\xi_j \in \mathcal{P}$ if j is even, and $\xi_0 = a_1$. Note that for a fixed positive integer k, system (16.9) leaves the subset defined by $\xi_j = 0$ for all $j \ge k+1$ invariant. On this invariant subset, system (16.9) becomes the following system for $(\xi_1, \cdots, \xi_k)$:

$$\begin{cases} (\xi_1)_{x_i} = [\xi_1, \ [a_i, \mathrm{ad}(a_1)^{-1}(\xi_1)]] + [\xi_2, a_i], \\ (\xi_2)_{x_i} = [\xi_2, \ [a_i, \mathrm{ad}(a_1)^{-1}(\xi_1)]] + [\xi_3, a_i], \\ \cdots \\ (\xi_k)_{x_i} = [\xi_k, \ [a_i, \mathrm{ad}(a_1)^{-1}(\xi_1)]]. \end{cases} \tag{16.10}$$

Thus we get

16.1 Theorem. *Let k be a positive integer, $a_1 + \sum_{j=1}^{k} \eta_j \lambda^{-j} = g(\lambda)^{-1} a_1 g(\lambda)$ for some $g \in \mathcal{F}$, $V_1 = \mathcal{K} \cap \mathcal{A}^{\perp}$, $V_j = \mathcal{K}$ if j is odd, and $V_j = \mathcal{P}$ if j is even. If $(\xi_1, \cdots, \xi_k) : R^n \to V_1 \times V_2 \times \cdots \times V_k$ is a solution of (16.10) with $\xi_0 = a_1$, $\xi_{k+1} = 0$, and $\xi_j(0, \cdots, 0) = \eta_j$ for $1 \le j \le k$, then $\mathrm{ad}(a_1)^{-1}(\xi_1)$ is a solution of the U/K-system (1.1) and $\mathrm{ad}(a_1)^{-1}(\xi_1) = g\sharp 0$.*

Figure 1

Example 10.8 (i). A 2-tuple in R^3 of type $O(2)$ obtained by applying a Ribaucour transformation to a pair of lines

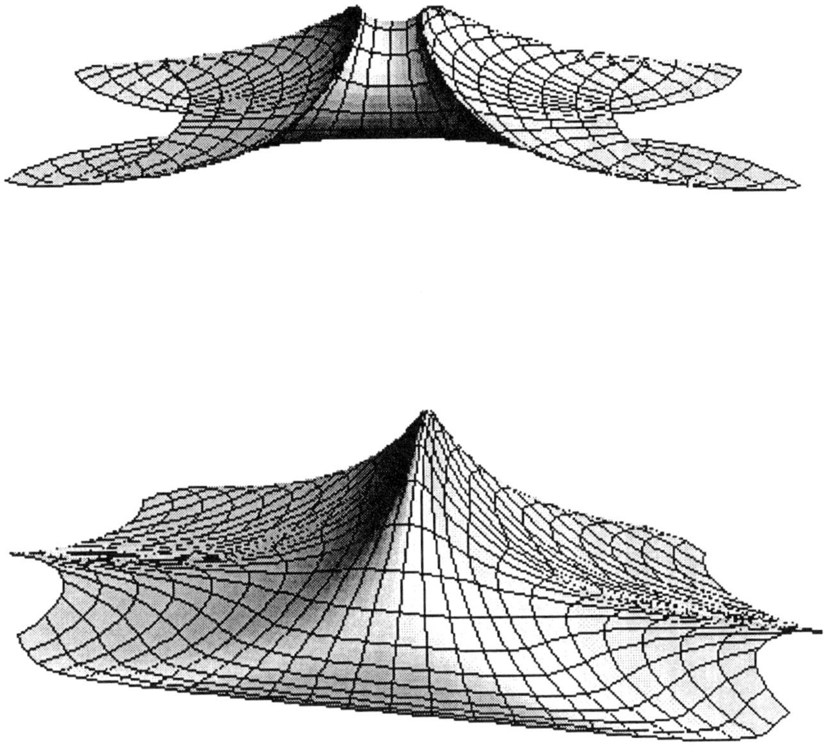

Figure 2

Example 10.8 (ii). A 2-tuple in R^3 of type $O(2)$ obtained by applying a Ribaucour transformation to a pair of lines

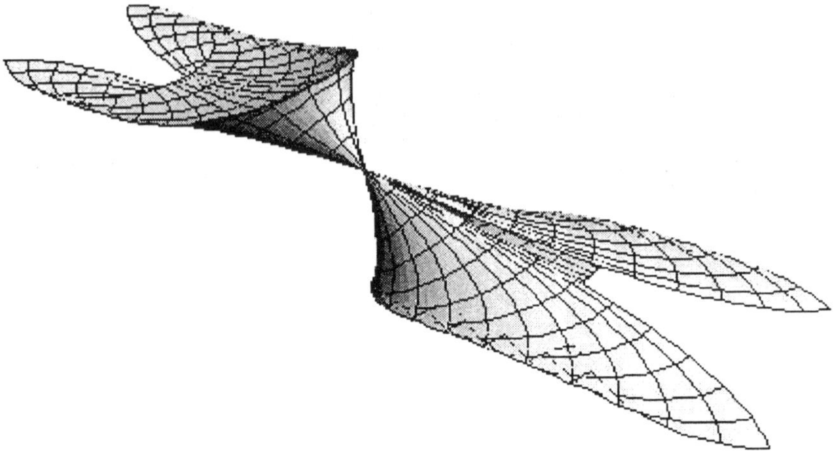

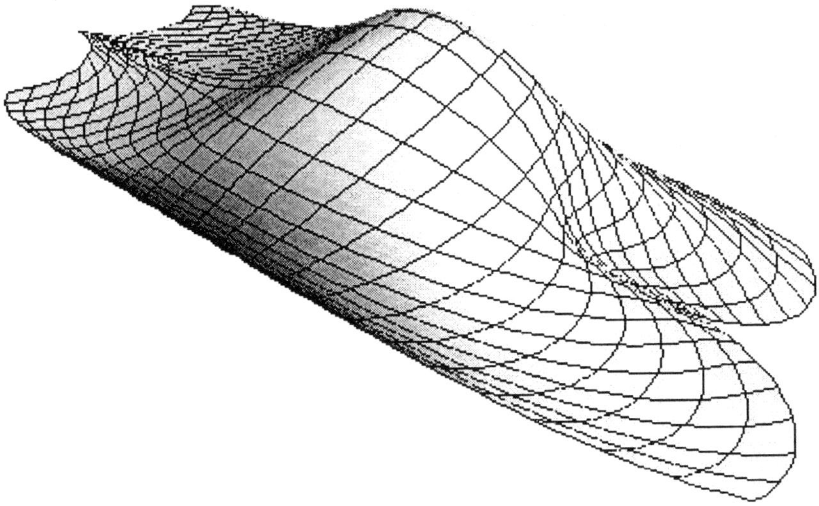

Figure 3

Example 10.10. $K = -1$ surfaces, top: case (1), bottom: case (2)

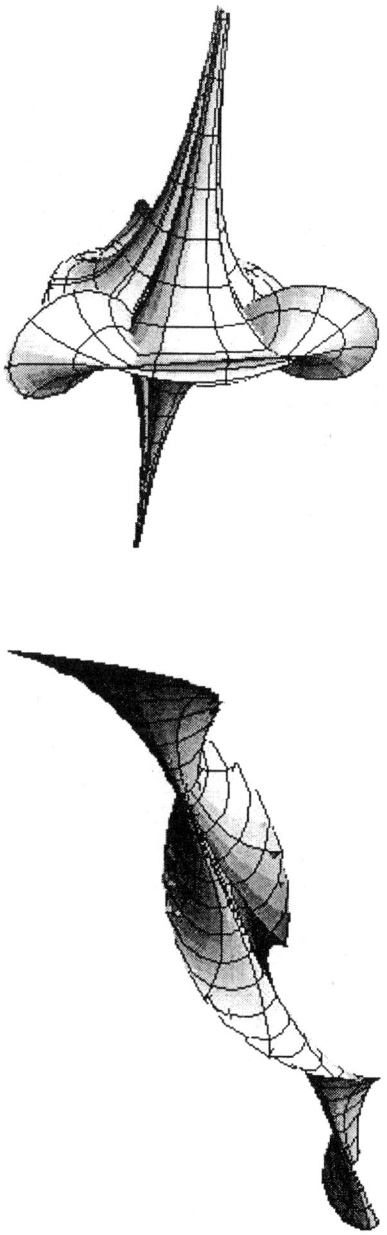

Figure 4

Example 13.2. An isothermic pair obtained by applying a Darboux transformation to the isothermic pair of planes

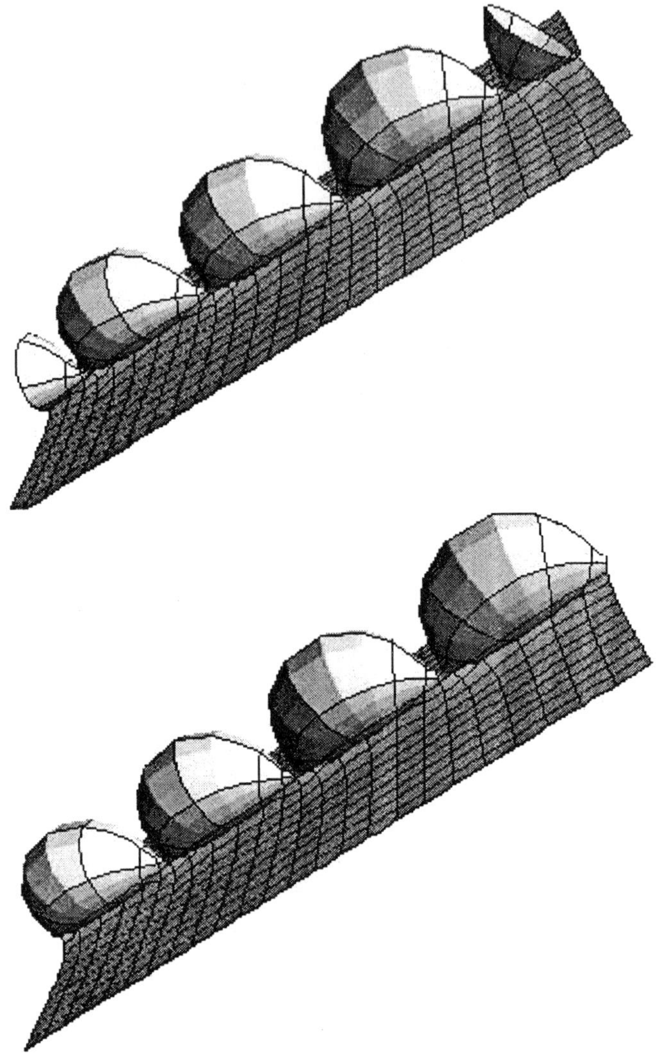

Figure 5

Example 13.5. A CMC surface obtained by applying a Darboux transformation to a cylinder
Top: surface $\tilde{Y}_1$ for $a = 2, \cosh c = 2$. Bottom: surface $\tilde{Y}_1$ for $a = 2, \cosh c = 4$.

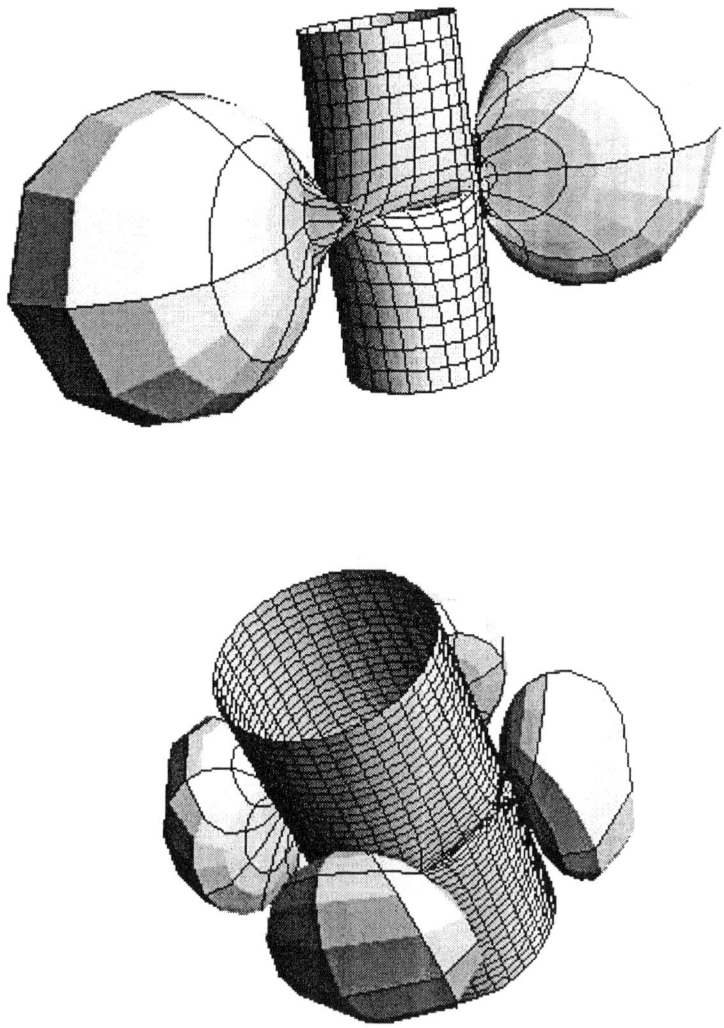

Figure 6

Example 13.5. A CMC surface obtained by applying a Darboux transformation to a cylinder

Top and bottom: two views of the surface $\tilde{Y}_1$ for $a = 2.75, \cosh c = \frac{9}{4}$

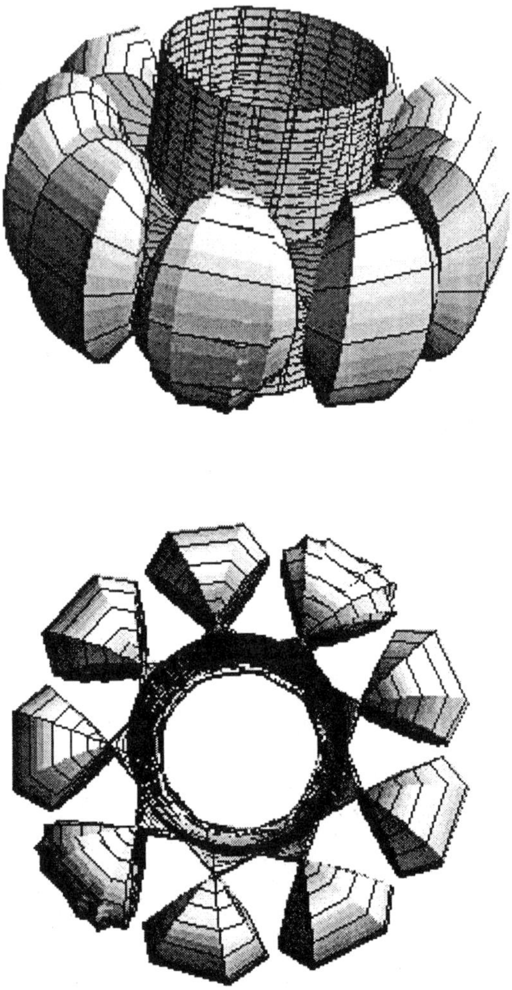

Figure 7

Example 13.5. A CMC surface obtained by applying a Darboux transformation to a cylinder

Top left and right are two views of the surface $\tilde{Y}_1$ for $a = 2, \cosh c = \frac{3}{2}$

Bottom left and right are two views of the surface $\tilde{Y}_1$ for $a = 2, \cosh c = \frac{7}{5}$.

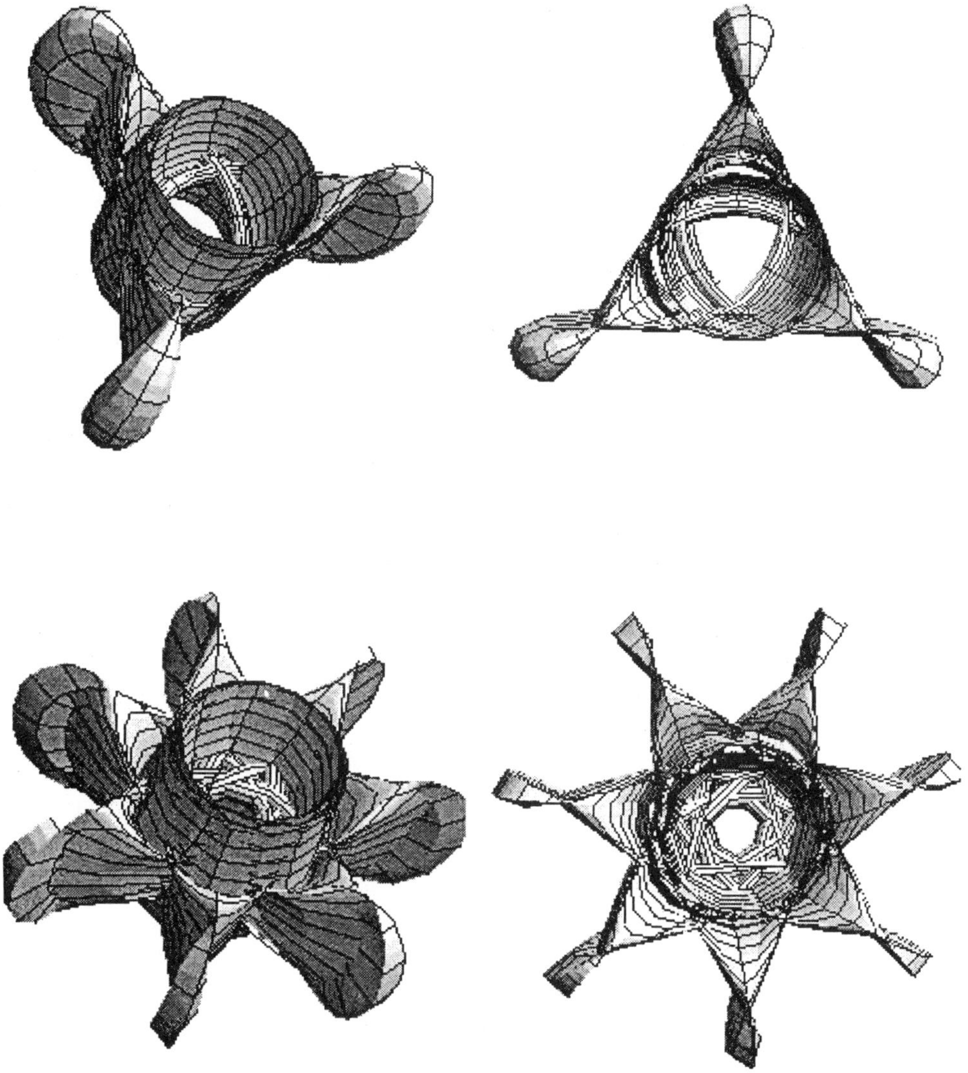

References

[Ba] Bäcklund, A.V., *Concerning surfaces with constant negative curvature.* New Era Printing Co., Lancaster, PA, 1905.

[BC] Beals, R., Coifman, R.R., Inverse scattering and evolution equations, *Comm. Pure Appl. Math.* **38** (1985), 39-90.

[Bi1] Bianchi, L., *Leconzioni di Geometria Differenziale.* Bologna, 1927.

[Bi2] Bianchi, L., Ricerce sulle superficie isoterme e sulla deformazione delle quadriche, *Annali Math. III* **11** (1905), 93-157.

[Bo1] Bobenko, A. I., All constant mean curvautre tori in R^3, S^3, H^3 in terms of theta-functions, *Math. Ann.* **290** (1991), 209-245.

[Bo2] Bobenko, A. I., Surfaces in terms of 2 by 2 matrices, old and new integrable cases, *Harmonic maps and Integrable systems, edited by A.P. Fordy and J. Wood* (1994), 83-127. Vieweg

[Bo3] Bobenko, A.I., Exploring surfaces through methods for the theory of integrable systems, lectures on the Bonnet porblem, math-dg/9909003

[BFPP] Burstall, F., Ferus, D., Pedit, F., Pinkall, U., Harmonic tori in symmetric spaces and commuting Hamiltonian systems on loop algebras, *Ann. of Math.* **138** (1993), 173-212.

[BHPP] Burstall, F., Hertrich-Jeromin, U., Pedit, F., Pinkall, U., Curved flats and isothermic surfaces, *Math. Z.* **225** (1997), 199-209.

[Bu] Burstall, F., Isothermic surfaces: conformal geometry, Clifford algebras and integrable systems, preprint, math-dg/0003096

[Ca] Cartan, E., Sur les variétés de courbure constante d'un espace euclidien ou non-euclidien, *Bull. Soc. Math. France* **47** (1919), 132-208.

[Ch] Cherednik, I.V., *Basic methods of soliton theory, Adv. ser. Math. Phys. v. 25.* World Scientific, 1996.

[CGS] Cieśliński, J., Goldstein, P., Sym, A., Isothermic surfaces in E^3 as soliton surfaces, *Phys. Lett. A* **205** (1995), 37-43.

[Ci] Cieśliński, J., The Darboux-Bianchi transformation for isothermic surfaces, *Differential Geom. Appl.* **7** (1997), 1-28.

[C] Chen, B., Surfaces with a parallel isoperimetric section, *Bull. Amer. Math. Soc.* **79** (1973), 599-600.

[DT] Dajczer, M., Tojeiro, R., Commuting Codazzi tensors and the Ribaucour transformation for submanifolds, preprint

[Da] Darboux, G., *Lecon sur la théorie générale des surfaces.* Chelsea, 1972. 3rd edition

[FP1] Ferus, D., Pedit, F., Curved flats in symmetric spaces, *Manuscripta Math.* **91** (1996), 445-454.

[FP2] Ferus, D., Pedit, F., Isometric immersions of space forms and soliton theory, *Math. Ann.* **305** (1996), 329-342.

[H] Helgason, S., *Differential Geometry, Lie Groups, and Symmetric Spaces.* Academic Press, 1978.

[HP] Hertrich-Jeromin, U., Pedit, F., Remarks on the Darboux transform of isothermic surfaces, *Doc. Math.* **2** (1997), 313-333.

[HMN] Hertrich-Jeromin, U., Musso, E., Nicolodi, L., Möbius geometry of surfaces of constant mean curvature 1 in hyperbolic space, preprint math/9810157

[M] Moore, J. D., Isometric immersions of space forms in space forms, *Pacific Jour. Math.* **40** (1972), 157-166.

[PiS] Pinkall. U., Stering, I., On the classification of constant mean curvature tori, *Ann. of Math.* **130** (1989), 407-451.

[PS] Pressley, A. and Segal, G. B., *Loop Groups.* Oxford Science Publ., Clarendon Press, Oxford, 1986.

[PT] Palais, R.S., Terng, C.L., *Critical Point Theory and Submanifold Geometry.* Lecture Notes in Math., Springer-verlag.

[Sa] Sattinger, D.H., Hamiltonian hierarchies on semi-simple Lie algebras, *Stud. Appl. Math.* **72** (1984), 65-86.

[Sy] Sym, A, Soliton surfaces, *lett. Nuovo Cim.* **33** (1982), 394-400.

[Ten] Tenenblat, K., Bäcklund's theorem for submanifolds of space forms and a generalized wave equation, *Boll. Soc. Brasil. Mat.* **16** (1985), 67-92.

[TT] Tenenblat, K., Terng, C.L., Bäcklund's theorem for n-dimensional submanifolds of R^{2n-1}, *Ann. Math.* **111** (1980), 477-490.

[Te1] Terng, C.L., A higher dimensional generalization of the sine-Gordon equation and its soliton theory, *Ann. Math.* **111** (1980), 491-510.

[Te2] Terng, C.L., Soliton equations and differential geometry, *J. Differential Geometry* **45** (1997), 407-445.

[TU1] Terng, C.L., Uhlenbeck, K., Poisson actions and scattering theory for integrable systems, *Surveys in Differential Geometry: Integrable systems (A supplement to J. Differential Geometry)* **4** (1998), 315-402. preprint dg-ga 9707004

[TU2] Terng, C.L., Uhlenbeck, K., Bäcklund transformations and loop group actions, *Comm. Pure. Appl. Math.* **53** (2000), 1-75.

[TU3] Terng, C.L., Uhlenbeck, K., Geometry of solitons, *Notices, A.M.S.* **47** (2000), 17-25.

[U] Uhlenbeck, K., Harmonic maps into Lie group (classical solutions of the Chiral model), *J. Differential Geometry* **30** (1989), 1-50.

[ZS] Zakharov, V.E., Shabat, A.B., Integration of non-linear equations of mathematical physics by the inverse scattering method, II, *Funct. Anal. Appl.* **13** (1979), 166-174.

[Zh] Zhou, Z.X., Darboux transformations for the twisted $so(p,q)$-system and local isometric immersions of space forms, *Inverse Problem* **14** (1998), 1353-1370.

Editorial Information

To be published in the *Memoirs*, a paper must be correct, new, nontrivial, and significant. Further, it must be well written and of interest to a substantial number of mathematicians. Piecemeal results, such as an inconclusive step toward an unproved major theorem or a minor variation on a known result, are in general not acceptable for publication. Papers appearing in *Memoirs* are generally longer than those appearing in *Transactions*, which shares the same editorial committee.

As of September 30, 2001, the backlog for this journal was approximately 7 volumes. This estimate is the result of dividing the number of manuscripts for this journal in the Providence office that have not yet gone to the printer on the above date by the average number of monographs per volume over the previous twelve months, reduced by the number of volumes published in four months (the time necessary for preparing a volume for the printer). (There are 6 volumes per year, each containing at least 4 numbers.)

A Consent to Publish and Copyright Agreement is required before a paper will be published in the *Memoirs*. After a paper is accepted for publication, the Providence office will send a Consent to Publish and Copyright Agreement to all authors of the paper. By submitting a paper to the *Memoirs*, authors certify that the results have not been submitted to nor are they under consideration for publication by another journal, conference proceedings, or similar publication.

Information for Authors

Memoirs are printed from camera copy fully prepared by the author. This means that the finished book will look exactly like the copy submitted.

The paper must contain a *descriptive title* and an *abstract* that summarizes the article in language suitable for workers in the general field (algebra, analysis, etc.). The *descriptive title* should be short, but informative; useless or vague phrases such as "some remarks about" or "concerning" should be avoided. The *abstract* should be at least one complete sentence, and at most 300 words. Included with the footnotes to the paper should be the 2000 *Mathematics Subject Classification* representing the primary and secondary subjects of the article. The classifications are accessible from `www.ams.org/msc/`. The list of classifications is also available in print starting with the 1999 annual index of *Mathematical Reviews*. The Mathematics Subject Classification footnote may be followed by a list of *key words and phrases* describing the subject matter of the article and taken from it. Journal abbreviations used in bibliographies are listed in the latest *Mathematical Reviews* annual index. The series abbreviations are also accessible from `www.ams.org/publications/`. To help in preparing and verifying references, the AMS offers MR Lookup, a Reference Tool for Linking, at `www.ams.org/mrlookup/`. When the manuscript is submitted, authors should supply the editor with electronic addresses if available. These will be printed after the postal address at the end of the article.

Electronically prepared manuscripts. The AMS encourages electronically prepared manuscripts, with a strong preference for $\mathcal{A}\kern-.1667em\lower.5ex\hbox{$\mathcal{M}$}\kern-.125em\mathcal{S}$-LaTeX. To this end, the Society has prepared $\mathcal{A}\kern-.1667em\lower.5ex\hbox{$\mathcal{M}$}\kern-.125em\mathcal{S}$-LaTeX author packages for each AMS publication. Author packages include instructions for preparing electronic manuscripts, the *AMS Author Handbook*, samples, and a style file that generates the particular design specifications of that publication series. Though $\mathcal{A}\kern-.1667em\lower.5ex\hbox{$\mathcal{M}$}\kern-.125em\mathcal{S}$-LaTeX is the highly preferred format of TeX, author packages are also available in $\mathcal{A}\kern-.1667em\lower.5ex\hbox{$\mathcal{M}$}\kern-.125em\mathcal{S}$-TeX.

Authors may retrieve an author package from e-MATH starting from `www.ams.org/tex/` or via FTP to `ftp.ams.org` (login as `anonymous`, enter username as password, and type `cd pub/author-info`). The *AMS Author Handbook* and the *Instruction Manual* are available in PDF format following the author packages link from `www.ams.org/tex/`. The author package can be obtained free of charge by sending email to `pub@ams.org` (Internet) or from the Publication Division, American Mathematical Society, P.O. Box 6248, Providence, RI 02940-6248. When requesting an author package, please specify $\mathcal{A}\mathcal{M}\mathcal{S}$-LaTeX or $\mathcal{A}\mathcal{M}\mathcal{S}$-TeX, Macintosh or IBM (3.5) format, and the publication in which your paper will appear. Please be sure to include your complete mailing address.

Sending electronic files. After acceptance, the source file(s) should be sent to the Providence office (this includes any TeX source file, any graphics files, and the DVI or PostScript file).

Before sending the source file, be sure you have proofread your paper carefully. The files you send must be the EXACT files used to generate the proof copy that was accepted for publication. For all publications, authors are required to send a printed copy of their paper, which exactly matches the copy approved for publication, along with any graphics that will appear in the paper.

TeX files may be submitted by email, FTP, or on diskette. The DVI file(s) and PostScript files should be submitted only by FTP or on diskette unless they are encoded properly to submit through email. (DVI files are binary and PostScript files tend to be very large.)

Electronically prepared manuscripts can be sent via email to `pub-submit@ams.org` (Internet). The subject line of the message should include the publication code to identify it as a Memoir. TeX source files, DVI files, and PostScript files can be transferred over the Internet by FTP to the Internet node `e-math.ams.org` (130.44.1.100).

Electronic graphics. Comprehensive instructions on preparing graphics are available at `www.ams.org/jourhtml/graphics.html`. A few of the major requirements are given here.

Submit files for graphics as EPS (Encapsulated PostScript) files. This includes graphics originated via a graphics application as well as scanned photographs or other computer-generated images. If this is not possible, TIFF files are acceptable as long as they can be opened in Adobe Photoshop or Illustrator. No matter what method was used to produce the graphic, it is necessary to provide a paper copy to the AMS.

Authors using graphics packages for the creation of electronic art should also avoid the use of any lines thinner than 0.5 points in width. Many graphics packages allow the user to specify a "hairline" for a very thin line. Hairlines often look acceptable when proofed on a typical laser printer. However, when produced on a high-resolution laser imagesetter, hairlines become nearly invisible and will be lost entirely in the final printing process.

Screens should be set to values between 15% and 85%. Screens which fall outside of this range are too light or too dark to print correctly. Variations of screens within a graphic should be no less than 10%.

Inquiries. Any inquiries concerning a paper that has been accepted for publication should be sent directly to the Electronic Prepress Department, American Mathematical Society, P. O. Box 6248, Providence, RI 02940-6248.

Editors

This journal is designed particularly for long research papers, normally at least 80 pages in length, and groups of cognate papers in pure and applied mathematics. Papers intended for publication in the *Memoirs* should be addressed to one of the following editors. In principle the Memoirs welcomes electronic submissions, and some of the editors, those whose names appear below with an asterisk (*), have indicated that they prefer them. However, editors reserve the right to request hard copies after papers have been submitted electronically. Authors are advised to make preliminary email inquiries to editors about whether they are likely to be able to handle submissions in a particular electronic form.

Algebra to KAREN E. SMITH, Department of Mathematics, University of Michigan, 525 University, Suite 2832, Ann Arbor, MI 48109-1109; email: `kesmith@math.lsa.umich.edu`

Algebraic geometry and commutative algebra to LAWRENCE EIN, Department of Mathematics, University of Illinois, 851 S. Morgan (M/C 249), Chicago, IL 60607-7045; email: `ein@uic.edu`

Algebraic topology and cohomology of groups to STEWART PRIDDY, Department of Mathematics, Northwestern University, 2033 Sheridan Road, Evanston, IL 60208-2730; email: `priddy@math.nwu.edu`

Combinatorics and Lie theory to SERGEY FOMIN, Department of Mathematics, University of Michigan, Ann Arbor, Michigan 48109-1109; email: `fomin@math.lsa.umich.edu`

Complex analysis and complex geometry to DUONG H. PHONG, Department of Mathematics, Columbia University, 2990 Broadway, New York, NY 10027-0029; email: `phong@math.columbia.edu`

***Differential geometry and global analysis** to LISA C. JEFFREY, Department of Mathematics, University of Toronto, 100 St. George St., Toronto, ON Canada M5S 3G3; email: `jeffrey@math.toronto.edu`

***Dynamical systems and ergodic theory** to ROBERT F. WILLIAMS, Department of Mathematics, University of Texas, Austin, Texas 78712-1082; email: `bob@math.utexas.edu`

Functional analysis and operator algebras to DAN VOICULESCU, Department of Mathematics, University of California, Berkeley, 970 Evans Hall, Floor 9, Berkeley, CA 94720-0001; email: `dvv@math.berkeley.edu`

Geometric topology, knot theory and hyperbolic geometry to ABIGAIL A. THOMPSON, Department of Mathematics, University of California, Davis, Davis, CA 95616-5224; email: `thompson@math.ucdavis.edu`

Harmonic analysis, representation theory, and Lie theory to ROBERT J. STANTON, Department of Mathematics, The Ohio State University, 231 West 18th Avenue, Columbus, OH 43210-1174; email: `stanton@math.ohio-state.edu`

***Logic** to THEODORE SLAMAN, Department of Mathematics, University of California, Berkeley, CA 94720-3840; email: `slaman@math.berkeley.edu`

Number theory to HAROLD G. DIAMOND, Department of Mathematics, University of Illinois, 1409 W. Green St., Urbana, IL 61801-2917; email: `diamond@math.uiuc.edu`

***Ordinary differential equations, partial differential equations, and applied mathematics** to PETER W. BATES, Department of Mathematics, Brigham Young University, 292 TMCB, Provo, UT 84602-1001; email: `peter@math.byu.edu`

***Probability and statistics** to KRZYSZTOF BURDZY, Department of Mathematics, University of Washington, Box 354350, Seattle, Washington 98195-4350; email: `burdzy@math.washington.edu`

***Real and harmonic analysis and geometric partial differential equations** to WILLIAM BECKNER, Department of Mathematics, University of Texas, Austin, TX 78712-1082; email: `beckner@math.utexas.edu`

All other communications to the editors should be addressed to the Managing Editor, WILLIAM BECKNER, Department of Mathematics, University of Texas, Austin, TX 78712-1082; email: `beckner@math.utexas.edu`.

Selected Titles in This Series

(Continued from the front of this publication)

704 **Jeff Hooper, Victor Snaith, and Min van Tran,** The second Chinburg conjecture for quaternion fields, 2000

703 **Erik Guentner, Nigel Higson, and Jody Trout,** Equivariant E-theory for C^*-algebras, 2000

702 **Ilijas Farah,** Analytic guotients: Theory of liftings for quotients over analytic ideals on the integers, 2000

701 **Paul Selick and Jie Wu,** On natural coalgebra decompositions of tensor algebras and loop suspensions, 2000

700 **Vicente Cortés,** A new construction of homogeneous quaternionic manifolds and related geometric structures, 2000

699 **Alexander Fel'shtyn,** Dynamical zeta functions, Nielsen theory and Reidemeister torsion, 2000

698 **Andrew R. Kustin,** Complexes associated to two vectors and a rectangular matrix, 2000

697 **Deguang Han and David R. Larson,** Frames, bases and group representations, 2000

696 **Donald J. Estep, Mats G. Larson, and Roy D. Williams,** Estimating the error of numerical solutions of systems of reaction-diffusion equations, 2000

695 **Vitaly Bergelson and Randall McCutcheon,** An ergodic IP polynomial Szemerédi theorem, 2000

694 **Alberto Bressan, Graziano Crasta, and Benedetto Piccoli,** Well-posedness of the Cauchy problem for $n \times n$ systems of conservation laws, 2000

693 **Doug Pickrell,** Invariant measures for unitary groups associated to Kac-Moody Lie algebras, 2000

692 **Mara D. Neusel,** Inverse invariant theory and Steenrod operations, 2000

691 **Bruce Hughes and Stratos Prassidis,** Control and relaxation over the circle, 2000

690 **Robert Rumely, Chi Fong Lau, and Robert Varley,** Existence of the sectional capacity, 2000

689 **M. A. Dickmann and F. Miraglia,** Special groups: Boolean-theoretic methods in the theory of quadratic forms, 2000

688 **Piotr Hajłasz and Pekka Koskela,** Sobolev met Poincaré, 2000

687 **Guy David and Stephen Semmes,** Uniform rectifiability and quasiminimizing sets of arbitrary codimension, 2000

686 **L. Gaunce Lewis, Jr.,** Splitting theorems for certain equivariant spectra, 2000

685 **Jean-Luc Joly, Guy Metivier, and Jeffrey Rauch,** Caustics for dissipative semilinear oscillations, 2000

684 **Harvey I. Blau, Bangteng Xu, Z. Arad, E. Fisman, V. Miloslavsky, and M. Muzychuk,** Homogeneous integral table algebras of degree three: A trilogy, 2000

683 **Serge Bouc,** Non-additive exact functors and tensor induction for Mackey functors, 2000

682 **Martin Majewski,** ational homotopical models and uniqueness, 2000

681 **David P. Blecher, Paul S. Muhly, and Vern I. Paulsen,** Categories of operator modules (Morita equivalence and projective modules, 2000

680 **Joachim Zacharias,** Continuous tensor products and Arveson's spectral C^*-algebras, 2000

679 **Y. A. Abramovich and A. K. Kitover,** Inverses of disjointness preserving operators, 2000

678 **Wilhelm Stannat,** The theory of generalized Dirichlet forms and its applications in analysis and stochastics, 1999

For a complete list of titles in this series, visit the
AMS Bookstore at **www.ams.org/bookstore/**.